Cases in Engineering Economy

SECOND EDITION

Cases in Engineering Economy

SECOND EDITION

William R. Peterson
Minnesota State University, Mankato

Ted G. Eschenbach
University of Alaska Anchorage

With cases contributed by

Kate Abel	Stevens Institute of Technology
E. R. "Bear" Baker, IV	University of Alaska Anchorage
Michael Dunn	Petrotechnical Resources of Alaska
Daniel Franchi	California Polytechnic State University – San Luis Obispo
Joseph Hartman	University of Florida
Paul Kauffmann	East Carolina University
Neal Lewis	University of Bridgeport
M. Lee McFarland	California Polytechnic State University – San Luis Obispo
Donald Merino	Stevens Institute of Technology
Karen Schmahl	Miami University
Herb Schroeder	University of Alaska Anchorage
Andrés Sousa-Poza	Old Dominion University
William Truran	Stevens Institute of Technology

Oxford University Press
New York   Oxford
2009

Oxford University Press

Oxford   New York
Auckland   Bangkok   Buenos Aires   Cape Town   Chennai
Dar es Salaam   Delhi   Hong Kong   Istanbul   Karachi   Kolkata
Kuala Lumpur   Madrid   Melbourne   Mexico City   Mumbai
Nairobi   São Paulo   Shanghai   Taipei   Tokyo   Toronto

Published by Oxford University Press, Inc.
198 Madison Avenue, New York, New York 10016
http://www.oup-usa.org

Oxford is a registered trademark of Oxford University Press

Library of Congress Cataloging-in-Publication Data

Eschenbach, Ted G.
   Cases in engineering economy / William R. Peterson & Ted G. Eschenbach. – 2nd ed.
      p. cm.
   ISBN 978-0-19-539783-3
   1. Engineering economy — case studies. I. Title.

TA177.4.E83   2009   658.1'52—dc19 88-22764

Printed in the United States of America

10 9 8 7 6 5 4 3 2 1

to

My wife Patricia,
Whose support and forbearance makes it all possible

Bill

My family,
Which models sensitivity, economy, teamwork, and engineering for me

Ted

ABOUT THE AUTHORS

William R. Peterson is an associate professor and chair of automotive and manufacturing engineering technology at Minnesota State University, Mankato. He received his PhD degree in industrial and systems engineering from The Ohio State University in 1995. He also holds an MBA from the Kearney State College (now the University of Nebraska-Kearney) and BIE from Auburn University in 1970.

Dr. Peterson has served on the faculties of Florida International University, Western Michigan University, Old Dominion University, and Arizona State University. He was the 3M-McKnight Distinguished Visiting Professor in Technology Development at the University of Minnesota Duluth. Prior to his doctorate, he worked in a wide variety of industries as an engineering and operations manager.

He is a past president of the American Society for Engineering Management; the Society for Engineering and Management Systems (of IIE); and Epsilon Mu Eta, the Engineering Management Honor Society; and a past chair of the Engineering Management Division of the American Society for Engineering Education. He serves on the editorial board of *The Engineering Economist.*

Ted G. Eschenbach, P.E. is a consultant and an emeritus professor of engineering management. He received his PhD degree in industrial engineering from Stanford University in 1975 and his MCE degree from the University of Alaska Anchorage in 1999.

Dr. Eschenbach developed the first edition of this casebook at the University of Alaska Anchorage, and he has authored or coauthored more than 10 engineering economy texts.

He is the founding editor emeritus of the *Engineering Management Journal.* He is a fellow and has served on the board of directors nationally for ASEM. He is a member of PMI and served for 20 years on the board of directors for the Anchorage chapter. He has served on the editorial board of *The Engineering Economist,* and he is now the area editor for case study analysis.

PREFACE

The case approach has long been a part of business, law, and medical education. Similarly, most engineering programs include design projects intended to bring real-world complexity into the classroom. Despite these traditions, many courses in engineering economy rely exclusively on end-of-chapter problems. Casebooks in engineering economy have not played the same role as similar books in finance, operations research, production, marketing, strategic management, etc.

End-of-chapter problems cannot illustrate the reality of economic analysis and managerial decision-making. They must simplify to explain basic techniques and principles. Once this foundation is understood, the student still faces the difficult transition to the chaotic, complex messes that characterize the real world. These cases step closer to reality. They also can be fun and motivating, as well as providing the opportunity for honing analytical, logical, and communicative skills.

While the casebook may be used to supplement *any* engineering economy textbook, it is available on the student and instructor CDs for engineering economy texts published by Oxford University Press.

Because it is designed as a supplemental case book, it does not explain the theory underlying the examples. It does provide an introductory chapter on case analysis, since this may be new to many students. A solved case is present in Chapter 2. Chapter 3 provides an overview of working in teams, since case analysis is often done in teams and many engineering programs do not include team training. Chapter 4 introduces sensitivity analysis because of its importance to case analysis and the weak coverage in some texts. Numerous cases have hints to the student. The solutions manual is available as an Excel file to adopters. The solutions manual includes tables identifying each case's topical coverage and corresponding chapters of leading textbooks.

The variety of cases and the options within them provide flexibility in the use of the casebook. It can be used at both undergraduate and graduate levels, as well as in professional seminars. It can be an integral part of a first course in engineering economy or of a course in capital budgeting, or it can be the text for an advanced case course.

For a first course in engineering economy this use can include both classroom examples and homework assignments. Reading of the cases outside of class can be efficiently combined with classroom discussion of difficulties, assumptions, relevant principles, and recommended approaches. Simple written assignments can include lists of assumptions and development of a basic cash-flow equation. Case usage can extend to detailed numerical solutions, comparisons of the results of different approaches, and complete written case reports. This option requires substantially more time, and it should probably be used on only a few cases per term. Courses in capital budgeting can use the casebook in a similar manner.

For subsequent courses the casebook could be used as the sole text with complete numerical analysis being supplemented with reading from the professional literature. At this level we suggest that the course organization should be case oriented.

For those that are familiar with casebooks for "capstone" courses we would like to emphasize that these cases are focused on applying the analytical techniques of engineering economy. Thus choices of quantitative data, assumptions, and tools play a larger role than the qualitative elements of more general business decisions. However, these qualitative elements are an essential part of case analysis, and they are included here.

We hope that this casebook is valuable to our colleagues in engineering economy, as we have relied heavily on their texts, their articles, and their ideas. We also hope that readers will help us improve later editions, by writing us with corrections, suggestions, complaints, or even encouragement.

We also hope that faculty will contribute class-tested cases of their own that can be considered for the next edition. Ted at aftge@uaa.alaska.edu will be the lead contact for this.

ACKNOWLEDGMENTS

These cases could not have been written and class-tested without the patience and understanding of our students. Especially Nick Kioutas of the University of Alabama in Huntsville and Mike Worden of the University of Alaska Anchorage who helped in preparing the instructor's manual. Many reviewers have helped strengthen the cases, and even more colleagues have encouraged us to complete this second edition. The largest contribution is from colleagues that have provided cases for this edition. These include:

Kate Abel of Stevens Institute of Technology
E. R. "Bear" Baker, IV of the University of Alaska Anchorage
Michael Dunn of Petrotechnical Resources of Alaska
Daniel Franchi of California Polytechnic State University – San Luis Obispo
Joseph Hartman of the University of Florida
Paul Kauffmann of East Carolina University
Neil Lewis of the University of Bridgeport
M. Lee McFarland of California Polytechnic State University – San Luis Obispo
Donald Merino of Stevens Institute of Technology
Karen Schmahl of Miami University
Herb Schroeder of the University of Alaska Anchorage
Andrés Sousa-Poza of Old Dominion University
William Truran of Stevens Institute of Technology

The draft manuscript for this edition was produced by Geri Dutton of Old Dominion University. Rachael Zimmermann, Patrick Lynch, Danielle Christensen, Peter Gordon, and Andrew Gyory of OUP were exemplary editors for this edition. This edition relied on the foundation of the first edition where Dundar Kocaoglu of Portland State University and Charity Robey at Wiley played key roles.

TABLE OF CONTENTS

Cases in Engineering Economy

SECOND EDITION

Chapter 1
Analyzing a Case

The major objective in using cases is to apply basic skills learned in doing end-of-chapter problems. This is done in realistic scenarios to foster critical thinking skills and to prepare students for the real world. Thus, cases must be very different from the end-of-chapter problems that typify homework. Thus, at first reading some students react with "What am I supposed to do?" or "There is no question being asked!" This chapter provides students with a starting point and outlines recommended approaches.

Cases lie along a continuum between end-of-chapter problems and the real world. End-of-chapter problems are well defined and generally the solution methodology is clear, even though it may require ingenuity and persistence for success. On the other hand, the real world has been described as a chaotic conglomeration of opinions, facts, and goals. Furthermore, the opinions, the goals, and even the facts may conflict or be contradictory.

Cases may report on a real situation as a newspaper or news magazine would, but more often the names are changed and the situation is simplified. Cases may also be based on a synthesis of multiple situations, set in a fictional description. It is the realism and the complexity of the scenario that defines a case.

Most students are comfortable with the clarity of end-of-chapter problems for homework, but the intrusion of real world complexity into cases causes student dissatisfaction. Cases may not have a clearly specified question or issue. Rarely is there a unique, correct answer. The data may be ambiguous, contradictory, missing, or irrelevant—like the real world. In addition there are difficulties in note taking and dependence on inefficient class discussion. The instructor may choose neither to direct class discussion nor to present solutions.

This chapter is a starting point in dealing with the messiness inherent in cases. The first step is to compare cases with end-of-chapter problems and the real world. The chapter then describes why cases are useful in learning and how they can be used. The last few sections detail a recommended series of steps for case analysis and for case presentation.

End-of-Chapter Problems vs. Cases

Differences between cases and end-of-chapter problems revolve around the following points:

- Ease of "topic" identification
- Data given and assumptions required
- Treatment of future uncertainty
- Emphasis on methodology vs. recommendation
- Role of non-quantifiable factors
- Complexity and length
- Need to use spreadsheets

Topic Identification. End-of-chapter problems are, by definition, linked with the chapter's topics. Often the problem order indicates which section covers the topic. On the other hand, cases almost always cover multiple topics from a number of chapters. The "main" topic might be apparent through the assignment's timing, but other topics can only be identified through careful reading.

Data and Assumptions. The student can expect that an end-of-chapter problem will contain all data required to solve it; and only occasionally will irrelevant information be included. Cases often have data that are incomplete, inconsistent, incorrect, or buried in irrelevant numbers and facts. Sorting out the relevant information depends on identifying the questions to be answered. Selecting the correct data from inconsistent possibilities may depend on perceiving the organizational biases and goals. Filling out the incomplete data requires a feeling for reality and identifying the root problem. Student assumptions may determine the conclusion, so that stimulating class discussions may focus on defending these assumptions. Thus, as in the real world, it is important to identify and document data sources and assumptions.

Uncertainty. Engineering economy focuses on how to base current decision making on future financial impacts. Unlike banking, which includes many fixed financial exchanges, virtually any engineering project involves significant uncertainty. For example, the first cost of a dam may be estimated within 20%, or the market demand for a new product may be estimated within [−50%, +300%].

For end-of-chapter problems with uncertainty, typically probability distributions are given, and expected values can be calculated. For cases, like many real world problems, prudent assumptions about the probability distributions are required. In addition with cases the treatment of uncertainty may require sensitivity analysis or scenarios of the best outcome, the most likely outcome, and the worst outcome. Recommendations will require judgment in evaluating the trade-offs between these possibilities.

Relative Importance of Methodology and Recommendations. Unlike more simple problems, cases emphasize decision making, not correctly choosing and using a formula to get the correct numerical answer. This requires integrating technical writing and presentation skills with analytical skills. The deliverables become a case report and presentation, rather than a worked out application of a formula.

In many cases the problem's definition is unclear and ambiguous, and reasonable objectives must be identified. For example, is it better to maximize the rate of return or to ensure a good rate of return with less risk? Even criteria may be unclear. For example, at what interest rate should the present worth be evaluated? Sometimes the objectives and criteria will be specified, and in other cases assumptions must be made and justified. Selecting objectives and criteria become the first steps in a methodological process that applies the tools of engineering economy.

Non-quantifiable Factors. The recommendations must often consider non-economic factors, such as organizational turf, personalities, or un-quantified growth potential. Some of these non-economic factors underlie the credibility of conflicting data, while others determine the objectives for individual projects, for organizational units, and for the entire organization.

Complexity and Length. Due to the details required for the previous points, a case is lengthier and more complex than an end-of-chapter problem. This difference means that cases are often approached differently than end-of-chapter problems. This chapter's final sections discuss a recommended approach.

Need to Use Spreadsheets. In engineering economy most end-of-chapter problems can be solved using formulas and the tables of engineering economy factors. Cases are almost always better solved using spreadsheets. Uncertainties in the data and the case's complexity usually require that the model be solved repeatedly as it is developed. Ensuring a model is correct is easier if you start with a simple model and add detail and complexity. Once it is developed, sensitivity analysis often requires multiple solutions. In addition, graphs are often needed to effectively support the recommendations. Thus spreadsheets are usually the best tool for case analysis—and for economic analysis in the real world.

It is recommended that these spreadsheets contain a data block where each element in the case is entered—once! It is also recommended that spreadsheet formulas make use of relative and absolute addresses to efficiently use the copy command for cash flow tables and for sensitivity analysis.

Cases vs. the Real World

These key points summarize the comparison of cases and reality:

- Reality accentuates the differences between cases and end-of-chapter problems.
- Reality presents the problem in more complex terms than do cases.
- Reality often requires iterative data gathering and investigation vs. fixed assumptions for cases.
- Reality increases the need for people skills.
- In the real world, analysis can threaten favorite projects or even jobs.

Reality vs. Cases vs. End-of-Chapter Problems. Cases and reality can be compared on the same seven points that were used to contrast cases with end-of-chapter problems. In these comparisons we find that reality simply is "somewhat more so" than cases. Just as cases are longer and more complex than end-of-chapter problems, reality is more complex than cases. The boundaries of real-world problems are less well defined than cases, partly because different problems overlap. Reality involves more uncertainty, a greater emphasis on the recommended decision, and a greater role for non-quantifiable factors. Reality is also far better addressed with spreadsheet models that are easy to modify and update.

Problem Presentation. The major difference between cases and reality is that cases present the problem to the student. In reality, the data must often be searched for, the objectives and

4

constraints must be identified, and the alternatives must be defined and perhaps designed. Then the economic evaluation can begin.

Iterative Data Gathering vs. Assumptions. Most of the time, real-world analysis permits repeated interaction. For example, the head of the design department may supply a view of the problem along with important data. Later, the analyst may find that these data conflict with facts supplied by another department. Rather than simply choosing a set of "facts" to believe, the step of asking for the data can be repeated. This process also allows the analyst to learn and to ask better questions on each visit. Cases do not allow this interaction, thus the student must instead substitute assumptions. This usually includes explaining why the assumptions were needed and justifying the particular choices that were made.

Documenting data sources for assumptions or adding a name/date block for data supplied by a person increases the credibility of the final recommendations. We'd like to thank the person who first advocated this to us: Karen Schmahl who used this approach to build trusted models in industry before she authored Cases 1 and 2.

People Skills. Interacting with others depends on people and communication skills. Coworkers, superiors, and subordinates contribute, clarify, and critique; so, good analytical skills are not enough. Group work with classmates on cases requires and develops similar skills, except that reality requires sensitive, active listening to ferret out and define the problem and the data. Finally, both case analysis and the real world do involve the communication of findings, and thus they depend on and develop presentation skills. In the real world multiple people from different parts of the organization will review the analysis and recommendations—not just the instructor.

Jeopardy. One of the most significant differences between case analysis and the real world is that classmates are not severely threatened by questions and analysis. On the other hand, employees are often influenced by the career implications of management decisions on which projects should be accepted and which should be rejected. Superiors, coworkers, subordinates, and even the analyst will see favorite projects or even their jobs threatened by recommendations based on economic analysis. Dealing with these situations requires both care and highly developed people skills.

Despite all of these differences, cases are clearly several steps closer to reality than end-of-chapter problems. The value of this closer approximation of reality is the subject of the next section.

Cases and Learning

The objectives for case analysis flow naturally from the previous comparisons of cases with end-of-chapter problems and with reality. The analysis, discussion, and presentation of cases relate to student learning in at least four ways:
- Development of intellectual skills
- Promotion of attitudinal change
- Broadening the student's knowledge of the world
- Motivation

Intellectual Skills. The goal of case analysis is learning how to apply textbook knowledge to real-world situations. This learning requires that students be *actively* involved, which the reading, listening, and lecturing of traditional classroom methods cannot accomplish. Lectures can effectively explain theory and principles, but cases are far better at developing judgment, insight, and critical thinking.

When case analysis successfully meets this objective, students are better able to: (1) diagnose problems, (2) evaluate data, (3) analyze complex situations, and (4) make convincing oral and written recommendations. The evaluation of data and the subsequent analysis depend heavily on the student's textbook knowledge of engineering economy. Cases then establish the importance of justifying assumptions, validating the model, and checking and modifying the results.

Attitudinal Change. Case analysis can support changes in student's attitudes. For example, students are forced to become more self-reliant rather than depending on the instructor or the back of the book. Cases also focus attention on the situation and a recommendation rather than on the analytical tools. Finally, the focus on the recommendations and the time pressure to meet deadlines ensure that quick-and-dirty techniques and solutions must sometime be accepted for cases—just as in the real world.

Real-World Knowledge. Cases describe situations that increase student awareness. The role-playing that results can assist in the transition from textbook problem solver to analyst to decision maker. For students this exposure to a variety of situations can even help to guide career choices.

Motivation. Many students are motivated to extraordinary efforts and learning in case exercises. For some it is interest in a real-world problem that seems less of an academic exercise. For others, the motivation comes from the competitiveness of inter-group rivalry or from the team spirit of a small group.

We believe that the importance of these objectives is high and increasing. More analytical techniques are being used in the economic justification of projects. Spreadsheets are used every day for economic analyses. This implies an ever-greater need to sharpen students' diagnostic skills for problem identification and technique selection.

Preparing a Case

How should **you** prepare a case? Case analysis is a process that involves identifiable stages and activities:

- Reading
- Identifying and modeling the problem
- Creating or identifying alternatives
- Evaluating the alternatives
- Fitting the model to the real world to support a recommendation

Reading. Gathering information to analyze a case must clearly be based on reading, as the case is presented in a written format. But, how is it done? We recommend that you read the case four times: (1) skim, (2) carefully with notes, (3) iteratively to clarify points, and (4) read to double-check.

The first time through establishes a framework to organize your thoughts. The second time is a careful reading with notes and highlighting for significant facts. After this reading you should be able to identify the issues, the alternatives, and the objectives. The third reading is piecemeal and iterative as you build your model (typically with a spreadsheet) and develop your recommendations. The fourth and final reading should be careful and complete. After you have done your analysis and developed your recommendations, you should double-check for a fact, opinion, or number that you overlooked or forgot about.

Identifying and Modeling the Problem. This chapter has stressed the fuzziness of case analysis and the need to identify the problem to be solved. This diagnostic stage is often approached through one of the following four "modeling" processes:

- A historical model compares the situation with the past. For example, should the company invest in this project when previous projects have averaged a 14% rate of return?
- A model based on analogy compares the situation with an example from the environment. For example, should our agency prioritize projects by ranking on benefit/cost ratios as other agencies do?
- A planning model may rely on comparisons with an ideal. For example, many textbooks suggest opportunity cost as a basis for selection of a minimum attractive rate of return.
- Finally we all have models of financial decision making from our "personal" experiences. For example, grandma has stressed the need to be frugal and to build a financial cushion for contingencies.

First, the overall objectives and the pieces of the problem must be identified—then the context that defines *who* you are in the case and *who* the audience is for *what* deliverables. Obviously different models will emphasize different features of the same situation. So, multiple views should be used rather than applying a single model.

Other dangers at the stage of problem identification include: (1) confusing symptoms with problems, (2) making premature evaluations, and (3) accepting everything as fact. The end result should be an explicit problem statement that identifies the significance of each piece of the problem—both short-run and long-run.

Creating or Identifying Alternatives. Finding alternative solutions is closely linked to defining the problem. Without alternatives, there is no choice and no decision to be made. The future may be terrible, but that alone does not define a problem to be solved. Sometimes the alternatives are clearly stated in the case, and sometimes alternatives must be developed with creativity and insight.

One of the most costly and common errors is to miss good potential alternatives. Often these rely on a "larger" view of the problem. For example, rather than improving packaging or inspection, quality might be improved through redesigning the product. Other times, the best alternative is to use the most cost effective elements of an alternative and omit what might be "gold-plating." A new factory layout might be able to cost effectively re-use some existing equipment, rather than replacing all of it.

Another common error is to assume that the identified problem has a single objective or a single cause. If there are multiple goals or multiple causes, then alternative solutions must deal with each, not with just one. For example, a public works project might be intended to provide a service, to alleviate unemployment, and to stimulate further development. Each of these objectives is related to a "problem," and these "problems" may have multiple causes. Designing and selecting the best alternative must consider all of these.

The cash flows for each alternative will be based on data or assumptions. Together with identification of their basis, they become part of our documented model.

Evaluating. Evaluation begins with "number crunching" the data and the model. Often this will require a feedback loop through earlier steps to find missing data or to improve an alternative. This iterative process is one reason that most cases are analyzed using spreadsheets.

One error is to depend on a single analytical tool or measure to evaluate the probable and possible outcomes of alternatives. These outcomes will almost always be multidimensional. Specifically, any recommendation must be *financially feasible in the short run* and *financially rewarding in the long run*. This focuses on a cash-flow analysis, by period and over the alternative's horizon. Even the question of long-run financial reward depends on selecting an objective. Possibilities include maximizing present worth, minimizing risk while meeting reasonable standards, maximizing the internal rate of return, or maximizing product diversification and market growth.

Other qualitative criteria include links to the organization's strategic goals, any differential market advantage, and ease of implementation. Implementation may depend on the limitations of personnel and control systems, the impact of motivation and morale, and the presence of contingency plans.

Checking the Model Against Reality. Engineering economy is based on mathematics and theory, but the real world is not. Moreover, any model is a simplification of reality. Just as the model is a "fit" to the real world, the model's recommendations often must in turn be adjusted. Besides assumptions that were not perfectly satisfied, there may be policy considerations that were simply omitted from the model. These differences are a key part of case analysis. What are the limits of the case analysis that was done?

Conclusion. All of these stages and activities are designed to produce a specific recommendation, along with the careful analysis and logical argument that support it. Part of the foundation will often be assumptions or inferences that must be justified. These assumptions should be made instead of recommending further analysis or further data gathering. These are feasible recommendations *only* when there is a multistage decision, or when a flexible alternative can be designed to allow future shifts in direction.

The process outlined above is work, but with the right attitude it can be fun as well. Pretend the case is real and play it like a game. It has a lot more possibilities than a homework problem with one right answer and only one approved method for finding it.

Case Discussions

Case discussions require more from you than does a lecture. You must listen much more closely, as points that are missed or not understood cannot be found later in the text. When the case is covered over more than one class period, the discussion may explain or define key assumptions that will be part of later analyses and presentations. Other elements that will improve the class discussions include:

- Allowing for special needs
- Preparing for class
- Managing the discussion
- Choosing what to say

Allowing for Special Needs. Discussing a case can be as simple as answering the instructor's questions or as involved as a one-hour team presentation. But, the active involvement of the entire class means that students must tell the instructor about any special needs. Students can stand and speak loudly for the hearing-impaired, and the instructor can make allowances for those who cannot speak easily.

Preparing for Class. To benefit from the discussion, be sure to fully prepare before class. This includes mastering the facts and analyzing each case before class. Students must attend the class regularly, because you cannot gain from a discussion you do not hear. Reading the text can sometimes make up for a missed lecture, but there is no substitute for the case discussion.

Managing the Discussion. Case discussions can degenerate into unproductive arguments if the students and the instructor do not maintain a constructive tone. The instructor's efforts to guide the case discussion must be tolerated, and classmates must be treated with respect—even during vehement disagreements. Sticking to the topic is essential. The discussion will flow more smoothly and be easier to follow without sudden topic changes, without rehashing, and without repeating previous comments—especially your own!

Choosing What to Say. The discussion will be more interesting if provocative, unconventional, or creative ideas are advanced. It will be more fun with appropriate humor. And it will be more productive if comments are brief and constructive. Even while you push your own ideas, you must listen to others, and you must be ready and willing to change your mind.

The case discussion should be based on the principles gleaned from the textbook and the facts of the case. Other sources include personal experiences, comparisons with previous cases, material from other courses, and library research. But whatever the source of the comments, themes should not be overplayed. Vary contributions and do not make the same points on each case. When the case describes a historical situation, hindsight may be unfair. Unforeseen events can cause the best decisions to have horrible consequences.

Oral Presentations

Mastering the art of case analysis requires effective oral and written presentations. The skills that you have honed in your English classes and in discussions with friends and family are the foundation for successful presentations. This short section cannot wave a magic wand to improve your presentations. This section can only remind you of what you already know about making a formal presentation.

When You Are the Presenter. You should not prepare and read a written text. Brief notes are better at balancing interest and liveliness with the need to stay on track. It may be worthwhile to memorize an attention-getting first sentence, but rehearsal is better than reading for the rest of your talk. Everyone is nervous before speaking. Yet, simply acting more confident can reduce the butterflies. Your audience wants you to succeed.

Some of us may be dull or awkward speakers, but if we are prepared we can still be effective and, with practice, we can improve. The simple steps of making eye contact, observing the group for feedback, varying our voice rate and tone, and moving and gesturing in moderation can increase our effectiveness. We may never be on television, but all of us will make presentations to our coworkers, our bosses, and our subordinates. These presentations will all have deadlines, they will all have limited time, they will all be simpler than a written report on the same subject, and they will benefit from simple visual aids. Thus, cases again prepare us for the real world.

Targeting Your Presentation. What is the purpose and context of your presentation? Are you presenting a factual analysis or "selling" your project? Has the audience read the case, prepared their own analysis, or read your report? Is each presentation of a different case or are several presentations on the same case (and are you the first or last of the several)? Answers to these questions allow you to focus your presentation so that you provide enough background and development to support your recommendations without boring your audience with what they already know.

The Written Case Report

Technical writing may be focused on presenting details, on convincing the reader that the choice is correct, or on motivating personnel to follow through. Effectiveness in every case requires a clear purpose, an identified audience, and an understanding of "who" is writing. Whether the report is a proposal, a memo with attachments, a technical report, or a completed form, a common process is used for:

- Structuring the report
- Writing vs. speaking
- Writing and attention to detail
- Outlining and writing the report

Structure the Report. The written report will identify issues and problems, it will analyze and evaluate the alternatives, and it will present specific recommendations. However, the structure of the report will often present these steps in a different order.

Case reports, like job-related reports, will be read by others who are familiar with the topic and who are focused on the decision to be made. These reports should begin with a summary of recommendations, so that readers can quickly find the most important points. Often these points will be included in a table or bulleted list, and then supported by a key chart.

The report's body should describe the background, problem statement, key assumptions, etc. Depending on the report's length and your instructor's preferences, the order and format may change. Possible additions include a letter of transmittal, a table of contents, a list of figures, and supporting appendixes for data and assumptions, detailed calculations, and sources of information.

Writing vs. Speaking. Like any other writing, you must balance belaboring the obvious against providing insufficient detail. This balance relies on an understanding of what your reader can be expected to know. But for any reader, their understanding will be enhanced by your use of simpler words and sentences. Writing does allow and requires more complexity than speaking. Speaking can rely on visual cues, on tone and intonation, and on interaction. Writing must rely on selection of words with precise nuances and on subordination and joining of ideas through more complex sentence structures. While fiction often uses a narrative style, technical reports generally should not.

Writing and Attention to Detail. The greater attention to vocabulary and structure that comes with writing is extended to concern with grammatical correctness, spelling, and clarity. Many readers will equate careless writing with careless work. A report with misspelled words, sentence fragments, overly long paragraphs, exaggerated conclusions, or an unclear style will simply be dismissed as confusing or unreliable. Similarly a report that is too short may be dismissed as shallow, or one that is too long as padded.

Outlining and Writing the Report. The foundation for a solid report is laid with the initial outline. This is the stage where you identify descriptive headings and the logical flow of ideas for paragraphs. Then deciding where to use and carefully designing figures, bulleted lists, and tables can dramatically improve the effectiveness of your writing. Then as you write, you can concentrate on clarity and appropriateness of tone (word selection, sentence structure, and transitions). This approach will minimize—but it cannot eliminate—the need for a cut-and-paste redrafting and for a revision of unclear sentences. All writing must be proofread multiple times. Finally, the image of professionalism can be enhanced with appropriate binding of the final report.

Conclusion

Case discussions stimulate thought and provide practice at communicating ideas. They will also make it clear that no one has all of the answers. In the classroom and as a manager it is wise not to assume a rigid position until a full range of views and information has been assembled.

Except for the most technique-oriented cases, even your instructor will not have all of the answers. Thus the case discussion will be more democratic and less prescriptive than typical lectures. Instead there will be a range of answers, often depending on the viewpoint and objectives of those involved. The choice will be between workable and non-workable solutions, rather than aiming for a single optimal solution.

Since there is not a single answer, the evaluation of your work will be based on:

- Care in use of facts and background knowledge
- Demonstrated ability to identify and state problems and issues clearly
- Use of appropriate analytical techniques and sound logical argument
- Consistency of analysis and recommendations
- The ability to formulate reasonable and feasible final recommendations

Cases represent learning by doing. By creating live and effective practice situations, cases extend the doing of homework problems that underlies engineering education. The learning is not a set of facts, but rather skill acquisition and development. Thus the learning is harder to "see" and to measure, but it is still very real. It is also fun.

For Further Reading

Edge, Alfred G. and Denis R. Coleman, *The Guide to Case Analysis and Reporting 4ᵗʰ*, System Logistics, 1986

An Example Case
Power to the People

This chapter is presented in three parts: the case statement, a description of the solution process, and an example case report. Chapter 4 on sensitivity analysis elaborates on the solution technique used here.

Chapter 2

Part A
The Case Statement

The Power Authority for Northcentral New Hampshire (NcNH) faces increasing demands for electricity, and its aging coal-based facility must soon be replaced. There are three options: a hydroelectric project, a new thermal generation facility (coal, natural gas, or oil-fired), or buying their power from a larger, nearby utility.

The last option has been discarded. NcNH did not join the regional consortium of utility companies when it was formed thirty years ago to support construction of a large nuclear facility. Since then, the nuclear plant has suffered enormous cost overruns. Now that it has finally entered service, the rates of all of the utilities in the consortium are quite high. NcNH has emergency interties with other utilities, but the high wholesale cost of consortium power limits its use to emergencies only.

The shareholders' environmental group has been quite effective, partly because NcNH is a member owned cooperative. This first peak coincided with nationwide gains for the environmental movement, and it kept NcNH from joining the consortium for the nuclear facility. The cost savings of that decision, more recent concerns over global warming, and very effective leadership have increased the influence of the environmental group.

The environmental group is now lobbying hard for replacement of the aging coal facility and for a shift away from coal for the new facility. Ms. Black, the plant manager, and the engineers under her disagree. They support coal as the fuel of choice for the potential new facility. NcNH's general manager, Mr. Herbert, suspects that the environmental group will prevail; thus, he has asked the engineering group to analyze a petroleum based facility as well as a small hydroelectric dam that has been proposed.

There are no local petroleum deposits; thus, the world market price and supply is the critical determining factor. As other utilities in the regional consortium have thermal facilities that burn natural gas and fuel oil, Mr. Herbert anticipates no problem in arranging for either fuel. He also suspects that the difference between the two petroleum fuels will be much smaller than the difference between them and the dam. Thus he has ordered that the preliminary analysis be based on the dam and on natural gas.

From consultation with other utilities and information from an old engineering feasibility study, Mr. Herbert and Ms. Black have pulled together some rough estimates. For example, an old Corps of Engineers report describes a dam that would currently cost about $120 million to build and $6 million a year to operate. The Corps estimates that such dams last about 50 years, and this one should have a capacity of 1500×10^6 kilowatt-hours (kWh) per year.

Generation through natural gas turbines comes in much smaller increments of capacity, allowing estimated costs per kilowatt-hour to be used. At current prices and assuming average loads of 80% on individual generators, the manufacturer's data on production efficiency provide a cost estimate of $0.015/kWh. This includes amortization of the gas turbines, normal maintenance, and an allowance for major overhauls. It does not include an expected average cost increase of 1% per year due to differential inflation in the price of fuel.

NcNH currently generates 600×10^6 kWh/year and expects this to grow 4% per year. (This implies that the dam cannot fully meet the demand for its 50-year life.) The state's Public Utility Commission has historically limited NcNH to a 6% rate of return on its capital investment.

Ms. Black has assigned you the responsibility of analyzing the choice. She suggested that the first decision is to choose which sensitivity analyses should be conducted and how they will be presented to Mr. Herbert and the other managers.

The second stage is of course to analyze these two choices, make a recommendation, and support it. Ms. Black has emphasized that sensitivity analyses are crucial here, as this is really only the first step in a long process. Other alternatives will be developed and the data will be refined, but a complete set of analyses will help NcNH understand the possibilities.

Option. *Not considered for example solution, so can be used as basis for an assignment.*

In addition to the cost of power generation, a second consideration is the cost of peak capacity. Obviously, the generation of electricity varies over the day and over the year. For NcNH the daily peak is 1.9 times the daily average, while the low is only 0.5 of the average. Similarly, January has a peak of 1.8 times the annual average; July has a lower peak of 1.2 times the annual average; while April and September have seasonal lows of 0.6 and 0.7, respectively.

The ability of any dam to respond to this type of variability is limited by the seasonality of water flows and by the dam's designed capacity. These limits imply that the dam's theoretical capacity is two to three times larger than its "average" capacity of 1500×10^6 kWh/year.

Does consideration of peak capacity generation modify your recommendations?

Part B
The Solution Process

For simplicity, this process is described as a one pass, do-it-all-right-the-first-time process. No case analysis is that simple. Rather all analyses require that some parts be redone as your understanding increases.

Reading the Case

The first reading is to get an overall sense of the problem. This is to analyze several new power sources (natural gas and hydroelectric) with a focus on the cost of power. The second reading would involve highlighting of the various numbers that will be part of the economic model.

Identifying and Modeling the Problem

The problem is to build an economic model that allows sensitivity analyses for two alternatives. The utility's principal objective is to have low cost, reliable power. However, some stakeholders are more concerned with the environmental impacts. This is the first step in what will be a lengthy process where the alternatives will be repeatedly re-defined, so it is more important to develop a solid understanding of what the key variables are than it is to recommend a final choice.

The case limits are defined by Mr. Herbert, the general manager: compare the dam and natural gas alternatives. Ms. Black further emphasizes selecting which sensitivity analyses should be done and how they should be presented. The deliverables must satisfy these requirements.

Identifying the Alternatives

The choice is between two alternatives for replacing the aging coal-fired power plant. If demand grows rapidly, then a large hydroelectric investment may produce power very economically over the problem's 50-year horizon. However, if the growth is slower, then paying the high operating costs of gas-fired turbines is cheaper, because they can be incrementally added to match increases in power demand. Besides the uncertainty in the demand's growth rate, other base case assumptions can be challenged. For example, the horizon and discount rate are somewhat arbitrary selections, and a dam's initial cost is often much higher than expected. The cost of the turbine is stated per kilowatt hour. Thus, we can make the assumption that this cost is the default, do-nothing cost.

In the long run, other alternatives, such as a new coal facility and dams of different sizes will have to be considered. However, the given dam and gas turbine alternatives are a good starting point. Also in the long run, the environmental impacts of the different choices will play a large role.

Compiling the Data

Base Case Assumptions. The following variables are used in the economic model, and their values are found in the case.

Dam	First costs	$120 million
	Operating & maintenance (O&M) costs	$6 million/year
	Capacity	1500×10^6 kWh/year
	Life	50 years
Turbine	Turbine cost	$0.015 / kWh
	Growth rate for turbine	1%/year
General	Initial power demand	600×10^6 kWh/yr
	Demand's growth rat	4%/year
	Discount rate	6%/year

Because the cost of the turbine is stated per kilowatt hour, there is an implicit assumption that this cost is the default, do-nothing cost. Thus, it is natural to assume that incremental power will be generated by natural gas turbines if the dam's capacity is exceeded. Since demand over the dam's capacity will be met by gas turbines in either case, we can ignore demands over the dam's capacity.

A second assumption involves the treatment of salvage or residual values at the horizon. For simplicity, this analysis assumes that the dam and the turbine alternatives have similar residual values. Note that at a discount rate of 6%, taking the present worth of any difference at year 50 reduces the difference by a factor of 20, since $(P/F, 6\%, 50) = .0543$.

A third assumption is more arbitrary. When do the geometric gradients for demand and the cost of fuel start? Both geometric and arithmetic gradients are typically assumed to have no change or zero cash flow in period 1. Starting either gradient a year earlier would make the dam more attractive.

Summary of assumptions:

- Demand over dam's capacity met by gas turbines in either case ➔ ignore
- Residual values for dam and gas turbines about the same after 50 years
- Geometric gradients for demand and cost of fuel have no change in period 1

Limits of Reasonable Change. Ms. Black's instructions make it clear that we cannot assume that the data are completely accurate. But there is no information presented on how much the various elements might vary. We could try and research this on the Internet, or we can simply make reasonable assumptions based on some general principles. In general there seem to be more ways for things to go wrong than right, so cost over-runs can be much larger than cost under-runs. Also, the further into the future a cash flow occurs, the more uncertain it is likely to be.

The dam has four variables that define it economically: a first cost, an annual operating cost, a capacity, and a life. Construction costs for dams depend on "ground" conditions, labor relations, etc. But the variability is not balanced. Under runs are possible, but overruns are more likely, and they are apt to be larger. So a range of −40% to +100% is reasonable (or $72 million to $240 million from a base of $120 million). Narrower limits seem appropriate for operating costs (−40%, +60%) and capacity (−10%, +20%). The 50-year life has a "suspicious hint" of arbitrariness, so limits of 30 to 100 years will be used (−40%, +100%) for the sensitivity analysis.

The current operating cost of the turbine should be a fairly exact figure. The turbines are off-the-shelf manufactured items with known performance and cost characteristics. So the cost/kilowatt-hour need only be varied by (−20%, +20%). However, estimating the differential inflation rate for petroleum products over the next 50 years is very uncertain, so a much broader range is assumed for the inflation rate of the turbine's fuel (−80%, +200%).

There are two general economic parameters that are estimated: the annual growth rate in power demand and the discount rate for the analysis. Historically, forecasting the growth rate in demand for power has been very difficult. In fact, numerous nuclear power plants have been canceled due to demand that failed to materialize. Cost overruns and delays simply increase the cost of the power, while a flat demand curve has caused 60% complete plants to be dismantled or mothballed. Thus a range of −80% to +50% is used for the uncertainty in the demand's growth rate.

The discount rate is also somewhat uncertain. The rate for financing may be known, but a 6% discount rate may not adjust for the risk to the members of the cooperative or the opportunity cost of the capital invested. Using a range of 3% to 10% basically corresponds to −50% to +70%.

These limits can be summarized as:

		Lower limit	Upper limit
Dam	First cost	−40%	+100%
	Operating costs	−40%	+60%
	Capacity	−10%	+20%
	Life	−40%	+100%
Turbine	Cost/kWh	−20%	+20%
	Growth rate for turbine cost	−80%	+200%
General	Demand's growth rate	−80%	+50%
	Discount rate	−50%	+70%

Choice of Graphs. Graphs or tables of present worth versus each variable could easily be constructed, but with complex data it is usually easier to interpret a graph. Tables often just overwhelm people with lots of numbers. Rather than drawing separate plots of present worth for each variable, it is more useful to combine them to examine the relative sensitivity of the present worth to each variable.

Figure 2-1 of the case report is a tornado diagram that summarizes the impact of each variable on the present worth. This is the best graph for showing the relative sensitivity of the present worth to many variables. First the variables are ranked based on each variable's range of present worth values. Then the tornado diagram arrays them from the most to the least impact.

The four variables with the most impact are examined in more detail in a spiderplot (Figure 2-2). We could also select variables for a spiderplot by considering the variable's importance, uncertainty, and controllability.

Some pairs of variables may have interesting interactions or they may be linked by importance or logic. Figure 2-3 considers the interaction of life and discount rate, as these are left to the analyst or defined politically in a disconcerting number of cases. Other parameters are more often based on hard data or engineering estimates. Figure 2-4 considers the interaction of the two "engineering" variables that may individually result in the turbine having a lower cost than the dam.

Building the Model. The model is fairly basic cash-flow equivalence, except for the geometric gradients and the dam's capacity. The extensive sensitivity analysis in this case is easier to do with formulas based on equivalent discount rates, but a cash flow table is easier to present and understand. Doing both double-checks the answer of each.

$$\text{Dam PW} = \text{DamFirstCost} + [\text{DamO\&M} * (P/A, i, \text{life}]$$

$$\text{Turbine PW} = \text{PW Growth Phase} + \text{PW No-Growth Phase}$$

The demand and the cost of fuel for the turbine increase at constant rates rather than by a constant amount each year. Thus, this involves a geometric, not an arithmetic gradient. Either year-by-year entries in a cash flow table are required, or the geometric rates must be combined with the discount rate in equivalent rates that combine all relevant factors.

Given an initial power demand, a growth rate in demand, and the capacity of the dam; the year the dam's capacity is fully utilized ($N_{\text{at capacity}}$) can be calculated.

$$\text{Capacity} = \text{Initial demand} * (F/P, \text{growth rate}, N_{\text{at capacity}})$$

This year is equivalent to the end of the growth phase, since we have assumed that excess demand will be supplied by turbines.

Analyzing the Results. The various graphs confirm that the base case results and most variations favor the dam. However, the curves for three variables do cross over the breakeven line: demand rate, the dam's first cost, and the discount rate. Thus, these variables obviously merit further study. Also the variables will change simultaneously. We could create scenarios of sets of changes, but this would greatly expand the case. So only a limited comparison of two pairs is made (see Figures 2-3 and 2-4). Figure 2-4 clearly warrants more concern than does Figure 2-3, where fairly extreme changes are required. Basically, consider the "distance" between the base case and the breakeven line.

Breakeven analysis can facilitate considering non quantifiable factors, which will have to be included before a final decision is made in the future. If a project is close to breakeven, then the economics of the two alternatives do not differ significantly, and other factors dominate. If one alternative is clearly better economically, then the question is whether non-economic factors imply another choice is better.

Chapter 2

Part C
The Written Report

To: Ms. Black & Mr. Herbert
From: SCR (Student Consultants Rule)
About: Identifying key variables to compare a hydroelectric dam and natural gas turbines

Recommendations

1. The dam is the better alternative for this initial analysis.
2. This recommendation is most sensitive to changes in the dam's first cost, the growth rate in demand, and the discount rate.
3. Some mechanism needs to be developed for balancing political risks with each other and with the economics.

Discussion

The given data and reasonable assumptions about its variability are summarized below.

			Lower limit	Upper limit
Dam	First cost	$120 million	−40%	+100%
	Operating costs	$6 million/year	−40%	+60%
	Capacity	1500×10^6 kWh/year	−10%	+20%
	Life	50 years	−40%	+100%
Turbine	Cost/kWh	$0.015 / kWh	−20%	+20%
	Growth rate for turbine cost	1%/year	−80%	+200%
General	Initial power demand	600×10^6 kWh/yr		
	Demand's growth rate	4%/year	−80%	+50%
	Discount rate	6%/year	−50%	+70%

Additional assumptions are:

- Demand over dam's capacity met by gas turbines in either case ➜ ignore
- Residual values for dam and gas turbines about the same after 50 years
- Geometric gradients for demand and cost of fuel have no change in period 1

Only three of the top four variables in Figure 2-1 are "likely to change enough" to allow the turbine to be economically more favorable. Figure 2-2 can be used to estimate the breakeven values for these variables: 160% of the base case value for the dam's first cost,

140% of the base case value for the discount rate, and 40% of the base case value for the growth rate in demand.

Figure 2-3 shows that the preference for the dam is relatively insensitive to the value for its life, so long as that life is at least as long as the shortest life expected of 30 years. Figure 2-4 allows us to examine changes in the dam's first cost and the growth rate in demand at the same time.

However, there are very substantial risks associated with the dam. Basically, this alternative is inflexible, and the future is uncertain. And much of the uncertainty and downside risk is linked to these three variables. The environmentalists can delay the dam with lawsuits, which will increase costs. Also, projects of this size very often exceed their preliminary design estimates. The demand rate is very uncertain because it relies on population and industry estimates extending over a half century. And finally, NcNH's discount rate is fair, but it is open to attack by environmentalists who do not want to see the dam built. There are political risks with economic impacts for all power generation options.

The largest concern is the risk to current rate payers, management, and employees. If the dam is the best choice, it is because of efficiencies when the dam is operating at or near capacity—in twenty or more years. However, if there are overruns or delays that force NcNH to buy high-priced nuclear power, then all suffer in the next five to ten years.

Figure 2-1   Tornado Diagram for PW$_{cam}$ – PW$_{turbines}$

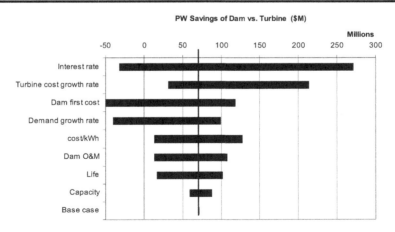

Figure 2-2   Spiderplot of Four Variables with Most Impact

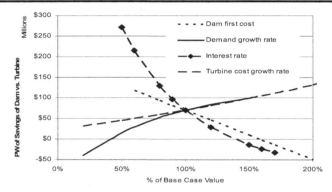

Figure 2-3   Discount Rate vs. Study Period

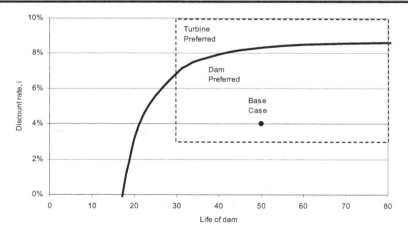

Figure 2-4   Demand's Growth Rate vs. Dam's First Cost

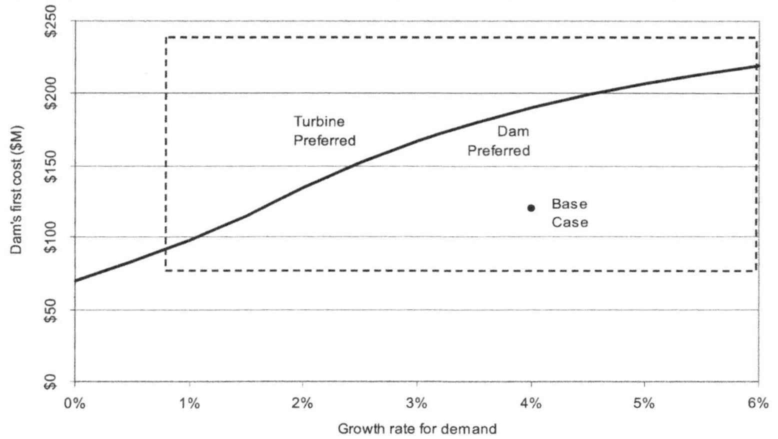

Appendix

The Problem

By direction of management this initial analysis is limited to the economic comparison of two alternatives – a hydroelectric dam and natural gas fired turbines.

The Data

The data summarized in the table is taken directly from the case. If there were different sources, then those sources would be identified here. The following table summarizes the basis for the limits on each variable.

Dam		Lower limit	Upper limit	Basis for limit
First cost	$120 million	−40%	+100%	Large overruns possible due to delays, ground conditions, etc.
Operating costs	$6 million/year	−40%	+60%	Over-runs likely to be larger than under-runs
Capacity	1500×10^6 kWh/year	−10%	+20%	Engineering design except for variability in water flow
Life	50 years	−40%	+100%	20 – 100 years seems more reasonable than exact 50 years
Turbine				
Cost/kWh	$0.015 / kWh	−20%	+20%	Known off-the-shelf technology
Growth rate for turbine cost	1%/year	−80%	+200%	Fuel costs are volatile
General				
Initial power demand	600×10^6 kWh/yr			Known value
Demand's growth rate	4%/year	−80%	+50%	Historically difficult to estimate
Discount rate	6%/year	−50%	+70%	Politically determined variable

The Model

The formula-based model is easier to use for sensitivity analysis, but the cash flow table model is easier to build and understand. Doing both double-checks the result.

In Figure 2-5, the years from 27 to 49 are hidden to help preserve readability. A data block in the top left corner of the spreadsheet defines *ALL* values in the spreadsheet. The PW equals CF_0 + NPV(interest rate, $CF_1:CF_{50}$) or =F17+NPV(A13,F18:F67).

Figure 2-5   Spreadsheet Model with Year-by-year Cash Flows

	A	B	C	D	E	F	G
				Plus/minus limits		Limits as factors	
1				Plus/minus limits		Limits as factors	
2	Dam			Lower limit	Upper limit	Lower limit	Upper limit
3	$120,000,000	first cost		-40%	100%	60%	200%
4	$6,000,000	O&M		-40%	60%	60%	160%
5	1,500,000,000	Capacity		-10%	20%	90%	120%
6	50	Life		-40%	100%	60%	200%
7	Turbine						
8	$0.015	cost/kWh		-20%	20%	80%	120%
9	1%	growth rate		-80%	300%	20%	400%
10	General						
11	600,000,000	Initial power demand					
12	4.0%	Demand growth rate		-100%	50%	0%	150%
13	6%	Interest rate		-50%	70%	50%	170%
14							
15						=NPV($A$13,F19:F68)+F18	
16					PW=	70,264,875	
17	demand	period	Dam	cost/kWh	Turbine	Dam-turbine	
18		0	-120,000,000			-120,000,000	
19	600,000,000	1	-6,000,000	$0.0150	-9,000,000	3,000,000	
20	624,000,000	2	-6,000,000	$0.0152	-9,453,600	3,453,600	
21	648,960,000	3	-6,000,000	$0.0153	-9,930,061	3,930,061	
22	674,918,400	4	-6,000,000	$0.0155	-10,430,537	4,430,537	
23	701,915,136	5	-6,000,000	$0.0156	-10,956,236	4,956,236	
24	729,991,741	6	-6,000,000	$0.0158	-11,508,430	5,508,430	
25	759,191,411	7	-6,000,000	$0.0159	-12,088,455	6,088,455	
26	789,559,068	8	-6,000,000	$0.0161	-12,697,713	6,697,713	
27	821,141,430	9	-6,000,000	$0.0162	-13,337,678	7,337,678	
28	853,987,087	10	-6,000,000	$0.0164	-14,009,897	8,009,897	
29	888,146,571	11	-6,000,000	$0.0166	-14,715,995	8,715,995	
30	923,672,434	12	-6,000,000	$0.0167	-15,457,681	9,457,681	
31	960,619,331	13	-6,000,000	$0.0169	-16,236,749	10,236,749	
32	999,044,104	14	-6,000,000	$0.0171	-17,055,081	11,055,081	
33	1,039,005,869	15	-6,000,000	$0.0172	-17,914,657	11,914,657	
34	1,080,566,103	16	-6,000,000	$0.0174	-18,817,556	12,817,556	
35	1,123,788,747	17	-6,000,000	$0.0176	-19,765,960	13,765,960	
36	1,168,740,297	18	-6,000,000	$0.0178	-20,762,165	14,762,165	
37	1,215,489,909	19	-6,000,000	$0.0179	-21,808,578	15,808,578	
38	1,264,109,506	20	-6,000,000	$0.0181	-22,907,730	16,907,730	
39	1,314,673,886	21	-6,000,000	$0.0183	-24,062,280	18,062,280	
40	1,367,260,841	22	-6,000,000	$0.0185	-25,275,019	19,275,019	
41	1,421,951,275	23	-6,000,000	$0.0187	-26,548,880	20,548,880	
42	1,478,829,326	24	-6,000,000	$0.0189	-27,886,943	21,886,943	
43	1,500,000,000	25	-6,000,000	$0.0190	-28,569,030	22,569,030	
44	1,500,000,000	26	-6,000,000	$0.0192	-28,854,720	22,854,720	
68	1,500,000,000	50	-6,000,000	$0.0244	-36,637,838	30,637,838	
69			=-$A$4		=-A68*D68		
70	=MIN(A67*(1+$A$12),$A$5)			=D67*(1+$A$9)			

Terminology for Formula-Based Model

i = basic discount rate

i_{eq} = equivalent discount rate

$$(1 + i_{eq}) = \frac{(1 + i)}{[(1 + \text{fuel growth rate}) * (1 + \text{demand growth})]}$$

Note 1: When i_{eq} would be negative, the right hand side is inverted.
Note 2: For the at-capacity phase, the (1 + demand growth rate) term is omitted for i_{eq} calculation.

LastGrowthYr = last year before capacity is reached

YrsAtCapacity = number of years with constant power usage

Mathematical Model

Dam PW   = DamFirstCost + [DamO&M * ($P/A, i$, life)]

Turbine PW = PW growth phase + PW at-capacity phase

                   = (1a) or (1b)   + (2a) or (2b)

a: if $i_{eq} \geq 0$

(1a) = TurbCostInitial * Init kWh * ($P/A, i_{eq}$, LastGrowthYr)

(2a) = TurbineCostLastGrowthYr * DamCapacity * (P/A, i_{eq}, YrsAtCapacity) *

                ($P/F, i$, LastGrowthYr)

b: if inversion is required to prevent $i_{eq} < 0$ (needed for tabulated factors, but not necessary if Excel functions, such as PV, are used)

(1b) = TurbCostInitial * Init kWh * ($1 + i_{eq}$) * ($F/A, i_{eq}$, LastGrowthYr)

(2b) = TurbineCostLastGrowthYr * DamCapacity * ($1 + i_{eq}$) * ($F/A, i_{eq}$, YrsAtCapacity) *

                ($P/F, i$, LastGrowthYr)

The spreadsheet in Figure 2-6 first builds the result for the $PW_{turbine-dam}$ piece-by-piece before they are combined using cut and paste. Also it is checked against the much easier to follow and verify result in Figure 2-5. This gives a single formula, which can be used to draw the spiderplot. The values at the bottom of the figure are used to draw the tornado diagram and the spiderplot. (A template for constructing tornado diagrams is included with the CD version of this casebook. For more explanation on constructing these figures, see chapter 4.)

Figure 2-6   Spreadsheet Model with Formula Basis

	A	B	C	D	E	F	G	H	I
1			Plus/minus limits		Limits as factors				
2	Dam		Lower limit	Upper limit	Lower limit	Upper limit			
3	$120,000,000	first cost	-40%	100%	60%	200%			
4	$6,000,000	O&M	-40%	60%	60%	160%			
5	1,500,000,000	Capacity	-10%	20%	90%	120%			
6	50	Life	-40%	100%	60%	200%			
7	Turbine								
8	$0.015	cost/kWh	-20%	20%	80%	120%			
9	1%	growth rate	-80%	200%	20%	300%			
10	General								
11	600,000,000	Initial power demand							
12	4.0%	Demand growth rate	-80%	50%	20%	150%			
13	6%	Interest rate	-50%	70%	50%	170%			
14									
15	-$214,571,164	PW dam							
16	24	last year dam not at capacity			=TRUNC(MIN(NPER($A$12,,$A$11,-$A$5)+1,$A$6))				
17	0.914%	Equivalent discount rate -- growth period			=(1+$A$13)/(1+$A$9)/(1+$A$12)-1				
18	4.950%	Equivalent discount rate -- at-capacity period			=(1+$A$13)/(1+$A$9)-1				
19	-$183,895,255	PW turbine growth phase			=PV($A$17,$A$16,$A$8*$A$11)/((1+$A$9)*(1+$A$12))				
20	-$408,702,636	FW(time 0 of at-capacity period)			=PV($A$18,$A$6-$A$16,$A$5*FV($A$9,$A$16,-$A$8))/(1+$A$9)				
21	0.2470	(P/F,i,last yr dam not at capacity)			=PV($A$13,$A$16,,-1)				
22	-$284,836,039	PW turbine			=A19+A20*A21				
23	$70,264,875	PW (dam - turbine)			=A15-A22				
24	$70,264,875	PW (dam - turbine)			Combined function built by cut & paste				
25	=-$A$3+PV($A$13,$A$6,$A$4)-(PV((1+$A$13)/(1+$A$9)/(1+$A$12)-1,TRUNC(MIN(NPER($A$12,,$A$11,-$A$5)+1,$A$6)),$A$8*$A$11)/((1+$A$9)*(1+$A$12))+PV((1+$A$13)/(1+$A$9)-1,$A$6-(TRUNC(MIN(NPER($A$12,,$A$11,-$A$5)+1,$A$6))),$A$5*FV($A$9,(TRUNC(MIN(NPER($A$12,,$A$11,-$A$5)+1,$A$6))),-$A$8))/(1+$A$9))*PV($A$13,TRUNC(MIN(NPER($A$12,,$A$11,-$A$5)+1,$A$6)),,-1))								
26									
27		Dam first cost	Dam O&M	Capacity	Life	cost/kWh	Turbine cost growth rate	Demand growth rate	Interest rate
28	20%						31333447	-28427635	
29	50%						44986451	14648389	271471939
30	60%	118264875	108093341		16821650		49779520	29147887	214631931
31	80%	94264875	89179108		49598458	13297668	59751489	52565898	129234516
32	90%	82264875	79721992	59130548	61121384	41781272	64938414	62016569	97108735
33	100%	70264875	70264875	70264875	70264875	70264875	70264875	70264875	70264875
34	120%	46264875	51350643	87620991	83263058	127232083	81353991	83976155	28677739
35	150%	10264875	22979294		94273543		99148054	99491461	-13079969
36	160%	-1735125	13622177		96530490		105408609		-23259003
37	170%	-13735125			98315228		111842493		-32069761
38	200%	-49735125			101721463		132240120		
39	300%						214038550		
40		Dam first cost	Dam O&M	Capacity	Life	cost/kWh	Turbine cost gr	Demand grov	Interest rate
41	min	-49735125	13622177	59130548	16821650	13297668	31333447	-28427635	-32069761
42	max	118264875	108093341	87620991	101721463	127232083	214038550	99491461	271471939
43	range	168000000	94571164	28490442	84899813	113934416	182705104	127919096	303541700

33

Chapter 3

Teamwork on Cases

While cases can be done on an individual student basis, often they are done by teams. The basic approach to solving the case remains the same as presented in Chapters 1 and 2:

- Reading
- Identifying and modeling the problem
- Creating or identifying alternatives
- Evaluating the alternatives
- Fitting the model to the real world

The difference is in the team's interactions as they accomplish the assignment.

Why Teams

Project analysis and justification in the real world is usually the result of a team effort. The different team members represent the project's different stakeholders and/or provide needed expertise on the project's different impacts on the organization—financial, production, people, market, and technological just to name a few. The quality of the final project and the extent of organizational agreement and commitment are often directly related to the variety of viewpoints considered in developing the project.

For example, if a new product is being considered, then marketing will need to estimate demand at various price levels and investigate the new product's impact on current products as well as how the new product fits into long-term marketing strategies.

If the project under consideration involves a cost reduction strategy such as automation, both human resources and production supervision are concerned. Production supervision needs to agree that the new staffing levels proposed will meet requirements and achieve projected savings. Human resources must develop a plan to reduce the number of departmental employees and ensure that all agreements (union contract and past-practice) are properly followed. If a new technology is involved, a group of technical experts may also need to be involved or developed.

If the product involves changes in the materials or in the way materials are supplied (for lean manufacturing or just-in-time production), material planning or purchasing needs to be involved. In automation projects, there are often material issues that need to be addressed. These issues may be as simple as which side of a box has the lap joint (which impacts how an automatic box erector functions) or as complicated as reduced tolerances on parts for automatic assembly.

Teams are a way of life in the practice of engineering. The ability to work productively in a team environment is a skill that all engineers are expected to develop. No single person will have all the skills and time to plan and implement a significant project. Almost all projects are done by teams of employees or even teams of employees, customers, and suppliers. In addition to the diverse skills required to develop, plan, and execute most projects, organizations have found that the decisions made by a team tend to be "better." With multiple inputs from diverse points of view, more alternatives are examined, while more potential benefits and costs (problems) are identified and addressed. From diversity comes a more holistic approach and solution.

The Case Team

Teams are groups of people working on a common problem with a common goal. The case is the common problem, and the common goal is its successful analysis. For the cases in this text, the recommended team is fairly small—three or four students (although teams of two to six are viable).

While teams can be for a single case, for a series of cases, or for the entire course, it takes time for a team's members to build a relationship with each other. If the team members already know each other, then the length of the "forming" period may be short. If some team members are unknown to each other, then we recommend spending some time just getting to know each other. If the team is to be successful, it must develop three key characteristics:

- A results orientation

- A focus on the efficient achievement of the desired result
- Trust and mutual support

Although listed last, trust and mutual support are of paramount importance in successful teams. The team members must be able to trust each other to complete assignments in a timely manner—at the time agreed to by the team and of a quality that the team can accept. The team must be able to trust its members to put forth their best individual efforts and to be working to complete the team assignment in a spirit of cooperation and fairness.

A results orientation is key for a successful team. The team is created to accomplish some specific goal. The achievement of this goal is the measure of team success. In practice and in the classroom, the team is successful only if it accomplishes its assignment. This is often emphasized with the aphorism, "there is no I in team." The team succeeds or fails as a team, and if it fails then no one succeeds.

This is the external measure of success. In solving cases, the desired result is fairly well defined although there is typically an open-ended aspect to cases. The required output or deliverable is typically a written report of "x" pages and an oral presentation of "y" minutes. What is needed to prepare these deliverables is the team's task. Developing and executing a plan to accomplish this task is the team's assignment.

If the external measure of success is achieved then (and only then) the internal measures of success come into consideration. These include: (a) Was the result achieved in an efficient manner? (b) Did the team stay on task while working on the assignment? (c) Did all the members contribute equitably to the deliverable?

To achieve both internal and external success, the team must organize itself to succeed. Each team needs to assign its members to the four roles needed for success. While each member may be assigned to multiple roles in any given team, each member's role needs to be clearly defined, and every team member must know who wears what hat on each case. If the team is going to exist for multiple cases, then the roles should rotate with each new case. The four fundamental roles in the team are:

- Team leader—This person coordinates the team, acts as the facilitator, and keeps the team on mission. Typically in the real world, this role is assigned or emerges as the team's activities progress. In this instance, selecting its leader is one of the team's first activities. It is important to note that this role is one of service to the team—not of one of authority. The team leader must facilitate the accomplishment of the team's assignment rather than simply direct the work of other members.

- Team scribe—This person records the team's plan of action, distributes it to the team members, and ensures that there is an agenda prepared for all team meetings.
- Team member—This includes everyone on the team, and the team members are tasked with accomplishing the task at hand.
- Team editor—This person takes the inputs from the team members (either as data or different sections written by different team members) and creates the final deliverable as a seamless artifact (report or presentation) for submittal.

Each member of the team may wear more than one hat. The team leader is also a team member. The team leader may also be the team scribe, although it is strongly suggested that these roles not be combined.

Two traits expected of each team member, regardless of role, are the ability to listen and the ability to contribute. Each team member must realize that the project is a team effort. To this end, each member must listen to the points offered by the other team members. Each member must also contribute their thoughts on the problem definition, the solution process (appropriate model), the possible alternatives, and the plan to accomplish the desired result. By balancing the trait of actively listening to the thoughts of others and the trait of presenting thoughts in a clear non-judgmental manner, each member can contribute to accomplishing the team's assignment effectively.

Team Activities

Each team is unique. Each case is different. This means that there is not one way for a team to interact or one way to structure a team's interaction. However, there are some general guidelines that will make the team activities less stressful and time consuming.

Team members must be prepared for each team activity. Whether the team meets face to face, in some virtual meeting format (conference calls, emails, bulletin board), or in a combination of formats, the key is to be prepared. To this end prior to team meetings there should be a formal agenda prepared and distributed beforehand. The team's first meeting should assign roles, set an agenda for the meeting, and agree to an approach to addressing the case. To be effective at this first meeting each member should have read the case multiple times.

Each meeting should result in a list of action items. Each action item should include the following detail:

- Who—the member (or members) responsible for this action item

- What—a description of what the action item is—preferably a tangible item or deliverable (an idea if articulated is a tangible output)
- By when—the latest date (and time on that date) that the promised "what" will be made available to the team by the "who" doing it

Each and every action item should have these three parts. Each action item must be agreed to by the team member to whom it is assigned. This buy-in is central to performance. It then becomes critical that the action be completed in a timely manner and to the level of quality expected—the best effort of the team member. Each team member must understand the significance of their action items to the project's successful completion. Each team member needs to be able to trust the others to meet the time committed to and to deliver a quality product as each item impacts the project's successful completion.

The product produced by the team should be homogeneous. While different parts can be produced by the individual team members, these parts must be blended into a single consistent report. The pronoun "we" not "I" should be used throughout. The style for the entire document should be consistent. The final product should then be read by each team member for correctness. It should be noted that everyone's style is a little different and the team members must keep this in mind when reading the final document—errors of fact and grammar must be corrected. However, avoid tinkering with the words just to achieve your version of perfection.

If the team needs to make an oral presentation, there are a few simple points to remember:

- Follow the same logical flow of a single speaker
- Let each speaker speak once during the presentation (do not have the same speaker bounce up and down in different portions of the presentation)
- Have each speaker present their contribution to the case analysis

By having each team member speak on their contribution, each team member is speaking on their area of expertise and it shows. By having each member speak once, each member's contribution to the whole is highlighted. By following the same flow as an individual speaker would in presenting the case, the team is presented as a unified entity, which reflects well on the members of the team.

Team Conflicts

It seems that conflict is inevitable in teams. Conflict can usually be traced back to one of several causes. The three most common causes of conflict in case teams are (1) failure to perform—not completing assignments on time, (2) failure to listen—insisting that one's way is the only way, and (3) a feeling that working in teams is not needed.

Each cause of conflict has two elements—the individual and team. The team must realize that each member is different and as such has different expectations and needs. The team member must realize that they must find a way to function in a team environment—most of their working career as an engineer will be spent working in teams, and their worth and often their salary will to a large extent be based on this ability.

When conflicts arise, and they will arise, the team should try to resolve the conflict internally. If this fails, the team (as a team) needs to request outside assistance from the instructor.

The abilities to compromise (not on principles but on opinions or approaches), to find win–win solutions, to persuade others to follow our lead, and to allow others to take the lead when appropriate are skills that will pay big dividends in your career. Conversely, the ability to know when help is needed is an important skill also. Remember—the *measure* of success is a deliverable that meets the need of the project (case).

Conclusion

In the workplace, teams are a common occurrence. The bigger and more important the project is; the higher the likelihood that a team will be used to do the project. Each team is unique, but successful teams are focused on results, and the members of a successful team have built a high degree of trust by meeting commitments, listening to the ideas of others, contributing their best ideas and efforts, and concentrating on the goal—a good product (be it a paper, a presentation, or leading a classroom discussion).

Chapter 4
Sensitivity Analysis

Uncertainty in the data is one of the largest differences between typical end-of-chapter problems and the more realistic problems of cases and the real world. While many management problems rely on quantitative data, rarely are those data exact. Estimation is required for some current values, and even more uncertainty is introduced by forecasting likely or possible future values. Engineering economy's emphasis on future cash flows and its application to projects at the preliminary design stage requires that this problem be specifically addressed.

Deterministic approximations ignore this uncertainty, which can be addressed through risk analysis, simulation, and sensitivity analysis. This last approach is often easiest, and it may also do the most to develop a "feel" for the problem. The simplest form of sensitivity analysis is to ask, "What if?" Different values for different scenarios can be entered into the spreadsheet to see how and if the recommended decision changes.

The next level of sensitivity analysis is examining the impact of reasonable changes in "base case" assumptions. As input to the sensitivity analysis, for each variable we need to know its most likely value, and lower and upper limits for reasonable changes in that value. From this we can calculate:

- The unit impact of these changes on the present worth (PW) or other measure of quality
- The maximum impact of each variable on the present worth
- The amount of change required to crossover a breakeven decision line

A sensitivity analysis can be done using any measure of quality, for example, injuries per year, net sales revenue, or internal rate of return. Present worth was used in Chapter 2 for the example case, and it is used here.

Breakeven Charts

A breakeven chart is constructed by holding all variables constant at their base case values and changing one variable to find its breakeven value, which is where the PW equals $0. This can be drawn without the lower and upper limits for reasonable change in the variable, but it is more useful if those are included. Figure 4-1 is an example. In this case, a heavy line weight is used for those values that are within the limits, and a light dashed line is used to extend the relationship to the breakeven value of $.0113/kWh. For graphing in a spreadsheet, this is done by using a second series of points for the heavy weight line.

Figure 4-1   Example Breakeven Chart for a Turbine's Cost per kWh

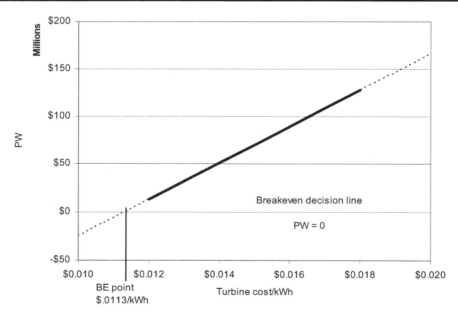

Relative Sensitivity Graph I – the Spiderplot

With a change in the *x*-axis, the information from a number of breakeven charts can be combined into a single relative sensitivity graph or spiderplot. For our example, this allows us to compare the relative sensitivity of the PW to changes in the growth rate in demand, the dam's first cost, the interest rate, etc. Because the "natural" units for these variables are all different (%, S, megawatt-hours/year), a common metric must be identified, for example, percentage of the base case value. (Note: This metric works poorly for variables with small or zero base case values. For these variables "natural" units work better.)

In the spiderplot, the variables are typically placed on the *x*-axis and the PW on the *y*-axis (see Figure 4-2). This orientation is chosen because the PW is a dependent variable.

Figure 4-2   Example Spiderplot

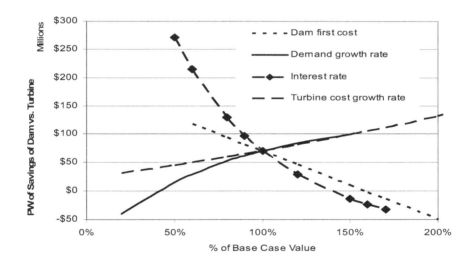

This *relative sensitivity graph* at its best allows the decision maker to quickly grasp:
- The range of reasonable values for each independent variable

- The unit impact of each independent variable on the PW (measured by the curve's slope)
- Breakeven values for variables whose curves cross the axis for PW = 0: 160% of the base case value for the dam's first cost, 140% of the base case value for the discount rate, and 40% of the base case value for the growth rate in demand

Combined these factors allow the decision maker to:

- Compare the maximum variation in PW for each variable
- Identify those changes that would support a different decision

At its worst, a relative sensitivity graph would mislead by distorting the scale to exaggerate or minimize slopes, by omitting any indication of the breakeven point, or by including too many or too few or the wrong variables.

The most common error is to graph all variables over the same range of percentage change. Imposing individual limits on the curves for each variable is crucial. Only then will this graph clearly describe the sensitivity of the PW to each variable. The heavy black line in Figure 4-1 corresponded to a ± 20% change in the cost per kWh for the turbine. If the curve for the turbine's cost were extended to ± 100% to match the *x*-axis limits of Figure 4-2, then the importance of this variable would be vastly overstated.

Just as a positive or negative PW cannot **determine** a decision by itself, but must instead be weighed against non-economic factors and the approximations of the model; so, too, must a breakeven curve or the crossing of PW = 0 be interpreted as a region of economic indifference.

Constructing a Spiderplot. If the *y*-axis value (PW in our example) can be expressed as a single formula, then this section describes how to build the table of values and draw the spiderplot. If the *y*-axis value can only be calculated using cash flow tables, then EXCEL DATA TABLES or PASTE SPECIAL/VALUES ONLY can be used to build the table of values and draw the spiderplot.

Even if the formula for the PW is very complex (cell A24 in Figure 2-6), the curve for each variable can be built by multiplying the variable's base case by a series of % values in the range between the minimum and maximum percentages of the base case. It is easier to build the table for all percentages of the base case and then to delete the "out-of-range" values. The cell addresses from Figure 2-6 are used as examples.

1. Create the data block of base case values and lower and upper limits. These limits are expressed as a % of the base case (E3:E13 and F3:F13).

2. Write the formula for the base case PW (cell A24) using absolute addresses (when the formula is copied, the addresses must not change).

3. Create the column headings (B27:I27) for the cash flow elements to be changed by copying from B3:B13 and using PASTE SPECIAL/TRANSPOSE. The row headings for the percentage points on the x-axis (A28:A39) are the different values listed in the lower and upper limits—listed in increasing order. Add a row for 100%—the base case.

4. Copy the formula for the base case into each cell of the top row of the table (B28:I28). Then edit the formula for *each* column, so that the cash flow element for that column is multiplied by the percentage value in column A. For example, *dam first cost* is the column head for B28:B39 and it is cell A3 in the data block. For row 28 in the *dam first cost* column every time $A$3 appears in the formula it is multiplied by cell A28. Repeat for the first formula in each column.

5. Copy the top row into the rest of the table. The values in the 100% row should all match the calculation of the base case PW. Check that the values are logical. For example does the PW go down as costs increase and go up as revenues increase?

6. Delete all entries that are beyond the limits for that column. For example, in this case the *dam's first cost* column should only have values between 60% and 200%. Similarly, the interest rate column should only have entries between 50% and 170%.

7. Create a graph by selecting the range A27:I39. The type should be XY. (If a line graph is selected, the x-axis values will be treated as labels and will be spaced evenly.) Each column of the table is a series. These series all have the same number of cells and *include the deleted entries*. For this example, the series for the dam's first cost is cells B28:B39, and the series for the x-axis is cells A28:A39.

8. If PW is the y-axis, add a variable that equals 0 for all x-axis values. If the y-axis is a B/C ratio, set the extra variable equal to 1. If the y-axis is the internal rate of return (IRR), set the extra variable equal to the minimum attractive rate of return. Add labels to the graph.

Typical Shapes to Expect. Understanding typical shapes for the curves for each variable is helpful in identifying when some kind of error has been made in building the economic model or in graphing the results.

For example, first costs, periodic payments or receipts, and other parameters found outside of the compound interest factors are usually linearly related to the PW. Variables—such as the discount rate, inflation and other geometric gradients, the life of a machine, or the problem's horizon—are found in the time value of money factors, and they exhibit a curved relationship to PW.

Stated graphically these curves are usually convex or concave. Stated economically, for investments, the discount rate has negative and decreasing returns to scale, and the project life has positive and decreasing returns to scale. For loans, the positive and negative flip-flop, but they both still exhibit decreasing returns to scale.

The large majority of variables show either a straight line or a concave or convex curved relationship to PW. So an expectation of linearity or decreasing returns to scale may be useful, but caution is necessary. Consider the dam's capacity in the example case. For considering the amount of power generated annually by the dam, this variable is outside of the compound interest factors. However, the magnitude of the dam's capacity also determines the number of years until this capacity is exceeded, which is inside some of the interest factors—so the net result is a complex curved relationship.

The curve for the PW, as a function of the demand's growth rate, is even more complex. (The demand's growth rate impacts the length of time until the dam's capacity is exceeded *and* the equivalent discount rate.) In examining this relationship, the smooth curve in Figure 4-3 was produced. The smoothing was, however, an artifact of the computer program being used. In Excel the choice is between a smooth curve and straight line segments. To produce the true curve with a sharp point, separate data series must be plotted on each side of that point with curve smoothing for each.

Figure 4-3   Dangers of Automatic Curve Smoothing

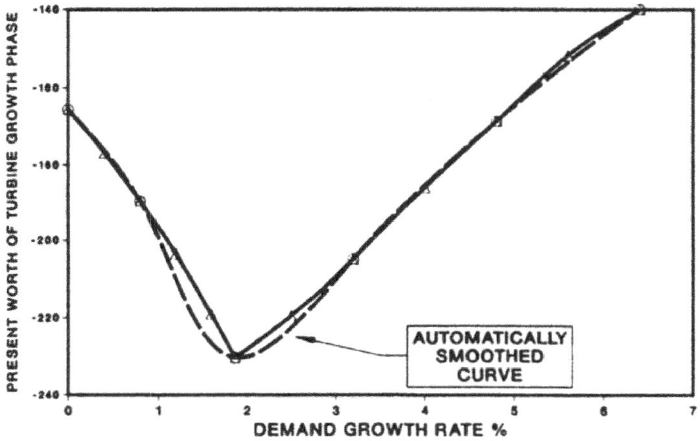

Relative Sensitivity Graph II – the Tornado Diagram

Economic models will often have far more variables than can be effectively shown on a spiderplot. Dozens of variables can be shown on a tornado diagram, such as Figure 4-4 from Chapter 2.

A tornado diagram does not show the actual relationship of the variables to the PW, only their possible impact on the PW. Thus, it doesn't show as much information, but it is easier to understand. It can also display the relative sensitivity of the PW to *all* variables. Figure 4-4 shows that three of the top four variables can result in a negative PW, which would possibly change the recommended decision.

Usually, but not always, the minimum and maximum values for the PW with each variable will occur at the limits for each variable. An example exception would be a firm operating at the optimal level, which is its maximum capacity. Then doing less would result in unsatisfied customers, and attempting to do more would fail and would create unsatisfied customers. For this example exception, the best PW would be from the "no change" percentage.

If a spiderplot is built, then adding rows for the minimum and maximum values for each variable is easy (A41:I42 in Figure 2-6). If no spiderplot is built, it is usually safe to assume that the minimum and maximum PW values occur at the extreme "allowed" values for each variable. In either case the range for each variable equals the maximum PW minus the minimum.

Then the tornado diagram is a stacked bar where each variable's bar begins at its minimum value and stacks the range to equal its maximum value. The TORNADO TEMPLATE included on the CD correctly accounts for the different possible combinations of negative and positive values. Since the tornado is normally displayed from most impact at the top to least impact at the bottom, these values must be sorted. Sometimes sorting with formulas creates errors, which can be avoided by using a copy with PASTE SPECIAL/VALUES to enter values into a copy of the template.

Figure 4-4   Tornado Diagram to Summarize the Relative Sensitivity of All Variables

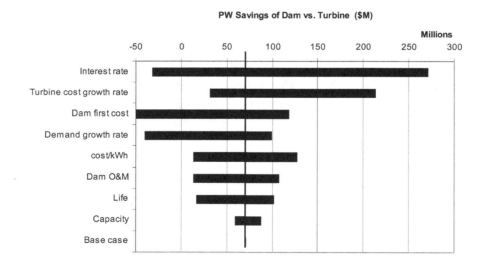

Multiple Variables or Multiple Alternatives

Sometimes it may be important to study the simultaneous impact of two variables on the decision to be made. For example, as shown Figure 4-4, the growth rate in demand and the dam's first cost have the most negative PWs in the example case. Other choices could be based on which variable has the largest slope or unit impact (the dam's first cost), the most uncertainty (the growth rate for fuel costs, −80% to +200%), or the greatest impact on the PW (interest rate or growth rate for turbine costs). It might also be important to study together variables such as the discount rate and study horizon that are often set by policy, or somewhat arbitrary selections of the analyst (see Figure 2-3).

Figure 4-5   Two-variable Breakeven Curve

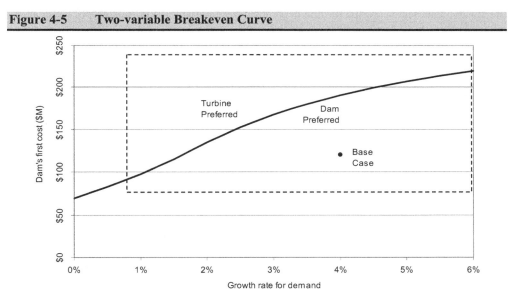

Computers can calculate tables of values that are "tiered" for any number of changing variables. However, only through limiting the number of variables to permit two-dimensional graphing is it easy to display the relationships. The breakeven or indifference line is where the PW of the proposal (go/no-go or option 1/option 2) is equal to zero. Any combination of the variables above the line will favor one option, while those below the line will favor the other option. This is shown in Figure 4-5. Note that the region within the identified lower and upper limits for each variable is shown with the dashed box.

This type of figure can also be constructed with the axes through the base case or with the axes measured as percentage change from the base. However, unlike the relative sensitivity charts, only one variable is being plotted along each axis. This allows the use of natural units, rather than percentage change for the axis variables. Since natural units are likely to fit the decision-maker's understanding better, they are a better choice when they can be used.

Figure 4-6 shows one way to display a sensitivity analysis for 3 variables on a 2-dimensional plot that is a series of breakeven curves. Obviously as the turbine cost increases, the dam cost can also increase before the turbine becomes the favored option. As the growth rate for demand decreases, the turbine becomes more favorable. With this graph, the analyst can determine how large a change in these three parameters can be sustained without changing the decision. If three variables are plotted, then the value of the third variable should be included in the annotation of the base case.

Figure 4-6   Multiple-variable Breakeven Curve

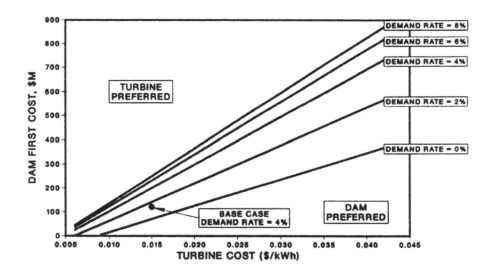

Similar plots with multiple alternatives, rather than multiple variables can also be constructed. Figure 4-7 is a more complex version of Chapter 2's Figure 2-4. This example adds a coal alternative with fixed and operating costs that are between the dam and turbine possibilities. Note that the breakeven line for coal and gas is vertical because this comparison is unaffected by the dam's first cost. In this type of graph each undominated alternative is identified with a portion of the feasible region.

Figure 4-7   Multiple Alternatives

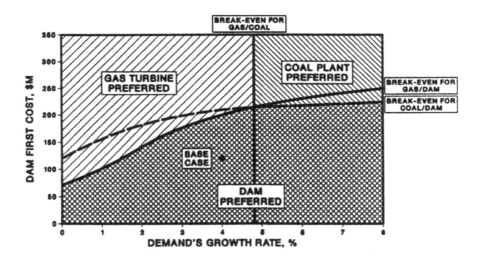

Summary

When constructing graphs to display sensitivity analysis, *all* of the graphs benefit from **clear delineation of the base case and the "breakeven" values**. *Breakeven charts* are best for analyzing one variable at a time. However for comparing variables, *spiderplots* (see Figure 4-2) can display for a limited number of variables:

1. The reasonable limits of change for each variable
2. The unit impact of these changes on the PW or other measure of quality

3. The maximum impact of each variable on the PW
4. The amount of change required to cross over the breakeven line

Tornado diagrams (see Figure 4-4) can display for all variables in a model:
1. The maximum impact of each variable on the PW
2. Which of these are enough to cross over the breakeven line

In case analysis and in applying engineering economy in the real world, uncertainty in a model's values is usually inescapable. Fortunately, spreadsheets give us the tools to efficiently and effectively describe the impact of this uncertainty.

For Further Reading

Eschenbach, T. G. and L. S. McKeague, "Exposition on Using Graphs for Sensitivity Analysis," *The Engineering Economist*, Vol. 37 No. 4, Summer 1989, pp. 315–333

Eschenbach, Ted, "Spiderplots vs. Tornado Diagrams for Sensitivity Analysis," *Interfaces*, Vol. 22 #6, Nov.-Dec. 1992, The Institute for Management Sciences, pp. 40–46

Eschenbach, Ted, *Engineering Economy: Applying Theory to Practice* 2nd, Oxford University Press, 2003, chapter 17

Eschenbach, Ted G., "Technical Note: Constructing Tornado Diagrams with Spreadsheets," *The Engineering Economist*, Vol. 51 No. 2, 2006, pp. 195–204

Tufte, Edward R., *Beautiful Evidence*, Graphics Press, 2006

New Office Equipment

by
Karen Schmahl
Miami University

You are one of the first engineers hired into the manufacturing engineering department at a brand new facility for producing aircraft engine controls. Your department is responsible for selecting, justifying, and installing all equipment for the factory. Most of the production equipment is being transferred up from the old facility in Fairfield, about a 3-hour drive away. The first production line has just been installed, and 8 more will be implemented over the next few months. When fully operational, approximately 700 production personnel will be employed on the factory floor. Most of your time is being spent on the shop floor learning about the equipment and trying to help with the installation process.

Right now most of the office cubicles are empty, but more people are being hired and transferring in from the main plant every week. Around 100 support personnel in operations management, quality engineering, manufacturing engineering, computer information systems, finance, purchasing, and human resources will have cubicle offices in a very large open area. Managers of each of the departments have already arrived. A separate design engineering liaison office will have approximately 20 personnel. The plant manager has a separate office suite.

Since the manufacturing engineering department is responsible for selecting and justifying all new equipment, it is being tasked with selecting and justifying the copy machines for the new office. For the transition period, two small copiers were purchased, one for the plant manager's office and another for receiving inspection. A few small copy machines were rented and placed around the office area, but now a decision must be made on equipment for full plant operation.

The manufacturing engineering manager has selected you for this project. In giving you this assignment, she tells you that she understands that it is not a very exciting piece of equipment for your first justification project. She explains that the task is a very important one, and it will give you the opportunity to meet personnel in all the other departments. It will also help you learn the steps of the machine selection and implementation process. The phone rings about then, and the manager is called to the shop floor to decide what to do with a machine that was damaged in transit. As she leaves, she asks you to start talking to people about their needs and to outline the steps you think would need to be taken to get this project done. She says she will review the outline with you tomorrow.

You should consider the following in preparing your outline:

1. How would you find out what the needs are for copy machines?
2. What are the project's objectives? Would different people in the office have different objectives? Are any of these objectives conflicting?
3. How can you find out what type of equipment is available and what it costs?
4. What additional considerations would you have to make in deciding on equipment and where to physically locate it?
5. How will you develop, evaluate, and compare alternatives?
6. Who do you think would have to approve your proposal?
7. How do you think an order would actually get placed for the equipment?
8. What steps might need to be taken before the equipment actually arrives to prepare for installation?
9. When should you follow-up to see that the equipment is meeting the needs?
10. How do the defined steps fit into the framework of the engineering design process? Where do the tools of an economic analysis fit in?

Budgeting Issues

by
Karen Schmahl
Miami University

Your new job is being an engineering manager in a division of a large firm. One of your first responsibilities is to prepare and submit a budget for next year's departmental expenses. Included in the expense budget are office and maintenance supplies, travel, education, and hospitality. You look at last year's budget and see that $50,000 was requested but only $45,000 was approved. In looking at the expenditures to date, you notice that, with three months remaining in the fiscal year, you still have $20,000 left in the expense account. The budget request is due in two weeks. You decide to ask managers from other divisions how they go about preparing their budget submissions.

The manager at the Greenville Division tells you that he always submits 10% more than last year's request. She comments that they always cut you about 20%.

The manager at the Newtown Division tells you that he tries to figure out where he will be with his expenditures at the end of this fiscal year, then estimates 15% higher for the next year. He says his first year he really had to scramble to cover his expenses because he only asked for what he thought he needed and they gave him 15% less.

The Boston Division manager says that he has all his personnel submit a "wish list" of items to put into the expense budget. He puts in for the whole amount, but tells his employees not to expect everything on the list. He also tells you that he always makes sure to be slightly over budget in any given fiscal year so that he can justify increasing the budget for the following year. He emphasized that if you don't spend all that you are given in a budget year, that they will probably give you even less the following year.

Suggestions to the Student:

1. What additional information would you want to have before preparing your budget submission?

2. What "strategies" could you take in submitting the budget?

3. Which strategies do you feel will most likely result in getting the amount of money that you need to run your department the following year?

4. What are the ethical implications of each of the strategies?

Case 3

Wildcat Oil in Kasakstan

by
Herb Schroeder
University of Alaska Anchorage

Wildcat Oil has recently discovered a new 500 million barrel crude oil reservoir in Kasakstan. Reservoir engineers predict recovery of about 300 million barrels with current technology. The firm needs a preliminary cost estimate for a feasibility study of a facility to produce the oil and prepare it for pipeline transmission. Wildcat has paid the Kasakstan government $400M in up-front lease costs. Additionally, the Kasakstan government will receive 10% of the net revenues (value/barrel [bbl] *minus* operating costs *minus* transportation costs). After 100 million barrels have been produced, all facilities and the remaining oil will belong to the Kasakstan government.

The feasibility study should optimize the trade-off between capital investment and production capacity in barrels/day (bbl/day). Engineering on a generic 36,000 bbl/day facility identified the major equipment items. Vendors have provided equipment costs for the five classes of major equipment (see Table 3-1) for this size facility. The factor estimates shown in Table 3-1 for all equipment, piping, and controls linked with each class of major equipment have been compiled from Wildcat's database based on past experience. For example, the total cost linked with the turbines is 2.5 times the $33.2 million (including the turbine cost).

Wildcat Oil uses a price of $19.50/bbl for oil of this quality delivered to the Kasakstan tanker facility. Facility operating costs are estimated at $4.50/bbl, and transportation to the tanker facility is estimated at $1.25/bbl. Production of all oil fields follows a decline curve;

however, negotiations between the government and Wildcat Oil have sized the facility so that production is basically constant through the period of Wildcat's ownership of the facility.

For estimating the cost of different size facilities, the production facility can be classed as a large refinery (with a power sizing or capacity exponent or Lang factor of .67).

Table 3-1   Cost Estimation Basis

Item	Cost (Millions of $)	Factor Estimates
Turbines	33.2	2.5
Compressors	24.8	2.8
Vessels & tanks	25.6	2.7
Valves	7.2	3.8
Switchgear	4.8	2.4

Suggestions for the Student:

1. What point of view should you take for analyzing this project?

2. How much should be budgeted for the 36,000 bbl/day production facility?

3. What additional costs and benefit(s), if any, are there to be derived from resizing the facility to process an additional 5,000 bbl/day?

Options:

1. What is the present worth of this project with a 36,000-bbl/day facility? An interest rate of 15% per year is appropriate for this type of investment.

2. Is the larger facility a wise investment?

Balder-Dash Inc.

by
Paul Kauffmann
East Carolina University

Balder-Dash Inc. (B-DI) is a large conglomerate with both manufacturing and service-based businesses. B-DI's goal is to provide its customers (who are original equipment manufacturers or OEMs) one-stop shopping for a wide range of component parts. B-DI's customers will pay a small premium for quality but are extremely cost conscious.

An opportunity to purchase an item it currently produces to sell as part of a bundle of products to one of its major customers has occurred. You have been assigned to recommend a course of action.

The item in question, a small fabricated metal part, currently is sold for $5.00 and has annual demand of 10,000 units. Demand is stable and reasonably uniform throughout the year. One of B-DI's suppliers was asked to quote the part and has come back with a price of $4.00. The current standard cost for this part is $6.70. The G&A (general and administrative) and R&D (research and development) costs add 20% to the standard cost for figuring profits on a per item basis. Accounting and purchasing have recommended discontinuing the part since no profit is made on the part. Marketing has proposed that the part be outsourced, which will convert an annual loss of $30,400 into a small profit. Marketing has indicated that discontinuing the part will not directly impact sales of the remainder of the bundle to the major customer but has expressed a concern that while pricing this item, the customer would "shop around" for other items in the bundle.

The current cost accounting picture is shown in Table 4-1.

Table 4-1   Current Standard Cost for Manufactured Part

Cost item	Cost per Unit	Annual Total
Direct material	1.00	10,000
Direct labor	4.00	40,000
Indirect labor (supervision)	1.20	12,000
Indirect overhead (warehouse)	.50	5,000
Total manufacturing costs	6.70	67,000
G&A / R&D costs	1.34	13,400
Total revenues	5.00	50,000
Profit (loss)	(3.04)	(30,400)

The proposed cost accounting picture is shown in Table 4.2.

Table 4-2   Proposed Standard Cost for Purchased Part

Cost Item	Cost per Unit	Annual Total
Direct material	4.00	40,000
Direct labor	0.00	0
Indirect labor (supervision)	0.00	0
Indirect overhead	0.00	0
Total manufacturing costs	4.00	40,000
G&A / R&D costs	.80	8,000
Total revenues	5.00	50,000
Profit (loss)	.20	2,000

The Manufacturing Process

Manufacturing takes place in a 1000 square foot area in an under-utilized warehouse on the main plant site. With improvements to planning and scheduling, lead times are dropping as are inventory requirements. It is projected that the raw material inventory will increase slightly over the next few years due to increases in demand, the work-in-process inventory will be stable, and the finished goods inventories will decline. Total warehousing requirements are expected to remain stable over the next five years.

The plant uses an overhead allocation of $5.00 per square foot per year. This is based on the plant's electric bill of $500,000; other utilities at $150,000; and building and grounds, maintenance, janitorial, and so forth at another $850,000. There are 300,000 square feet on the site, so the allocated cost is $1,500,000 over 300,000 square feet, which equals $5 per square foot per year.

The supervisor assigned to this product supports this product in addition to other duties in the main production facility. These duties include scheduling, maintenance oversight, as well as direction of direct labor for one of the plant's two main production departments—machining operations. When the part in question is being made, which is infrequent, this supervisor will visit the warehouse several times a day to assess the status of production. Currently one-fifth of his full time salary of $60,000 per year (including labor-related overhead) is allocated to this product.

As stated previously, this part is produced infrequently. When production of the part is in process, "production associates" from the "floater pool" (a group of cross-trained workers who fill in for vacations, sickness, and worker absences) are used. If five of the production associates in the labor pool are available, average weekly demand can be satisfied in one day. If fewer than five production associates are available, production takes correspondingly longer. Annual direct labor requirements for this part thus equate to one full time employee or $40,000 per year (including labor-related overhead).

The raw material for the part in question costs $1.00/kg if purchased new. However, the raw material for this part is not purchased directly but is the recovered drop from the production of another part (XV12C). The part in question consumes about half the drops from current production levels for XV12C. The drops not used in making the part in question are sold as scrap for $.25/kg (each drop weighs .75 kg). Currently, XV12C production is forecast to remain at current levels for the next three years.

G&A and R&D are allocated based on 20% of the cost of goods manufactured.

Suggestions for the Student:

1. Are there any inaccuracies in the standard costs?

2. Should the part in question be outsourced?

3. What is the financial impact of your recommendation?

4. What other recommendations (or investigations) would you make (or start) based on your findings to date?

Can Cruncher

One project in a junior mechanical engineering course was to design a manually operated aluminum-can compactor. The prototype in Figure 5-1 was developed and built by a team of three—Joyce, Ted, and Will. Having been convinced by their classmates that it would sell well, they are now trying to convince Ted's father to manufacture it in his machine shop.

This machine shop employs 15 people doing a large variety of small batch fabrications for about 40 local manufacturing plants. Recently, one of their two largest customers announced plans to move their plant to another state, thus the shop is facing a 20% drop in business.

Ted's father, Mr. Dick Strong, has identified three options for dealing with this loss of business. Through a combination of price cuts and vigorous marketing, he can attract business away from his principal competitors. He expects that this strategy would lead to financial ruin, as he has historically worked hard at marketing and competitive pressures have kept his profit margins relatively low.

He finds the option of layoffs nearly as unattractive. Skilled machinists are hard to find, so any he lays off are unlikely to be available for rehire later. Also the damage to shop morale will be high, making it likely that additional employees might leave, thus making it more difficult to hire replacements. Layoffs would also damage his self-respect and reputation, as he has never laid someone off and he has stressed that Strong Metalworks is a "family."

Figure 5-1   Compactor concept and operation

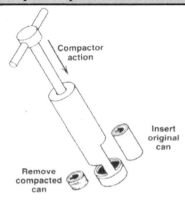

Figure 5-2   Aluminum can compactor design

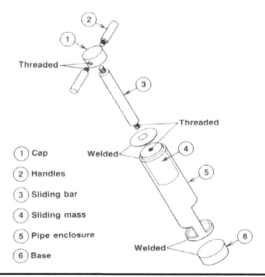

The only other option he can envision is adding a new kind of activity to his business. This could represent (1) a new process such as plating, which he now subcontracts; (2) a new market, such as sending his salespeople into the next city; or (3) developing products for wholesale or retail sale rather than simply fabricating components to be incorporated into the products of others.

Not surprisingly, Mr. Strong is most interested in this last possibility, and he is considering the suggested product from the student team. Although he has other possibilities, he would like to encourage his son and the other students. Whichever product is selected for development, if any, Mr. Strong recognizes that this initial venture into a new mode of operations is unlikely to replace the entire 20% drop in business. It may not even be profitable. Rather, he views this as an experiment whose primary value may be marketing information and experience. Whereas coping with the unanticipated decline in demand is the primary motivation, fabrication of complete products for direct sale will also allow the demand for labor to be smoothed through production to inventory.

Figure 5-1 shows the concept and operation of the student's can compactor. Figure 5-2 provides a visual guide to product configuration and assembly. Both drawings illustrate that the compactor operation consists of a sliding mass within a cylinder to compress the aluminum can. The window in the cylinder allows can insertion, while a second smaller window permits safe, automatic removal (it falls out).

In Table 5-1, the student time and material cost for product development are shown. The development has included a design effort, prototype testing, and final write-up.

After discussing the manufacturing process with Mr. Strong, the students agreed that the product design could be refined. This redesign will emphasize improving the can compactor's "manufacturability" and its effectiveness. Table 5-2 summarizes the expected time and material cost to modify the design and to produce 20 revised prototypes. Once the design is finalized, it is expected that the material cost can be trimmed by 40% and the manufacturing time by 70%. Although new capital equipment may be justifiable, Mr. Strong has ruled it out for now. Student time may be treated as a freebie. (They get learning, experience, and prototypes for themselves and friends.) The employees of Strong Metalworks average $17.50/hour in wages and fringe benefits. The burden rate for calculating the cost of administrative overhead is 55%. Mr. Strong has stated that this can be ignored for the redesign phase since those units will not be sold.

Table 5-1		Can Compactor Development Costs
Material $	Labor Hours	Item Description
	5	Generation of design alternatives
	8	Initial design and alternative selection
	18	Experimentation and design finalization
$22	18	Construction of prototype
$7	12	Production documentation

Table 5-2		Redesign and Initial Production
Material $	Labor Hours	Item Description
	8	Student team contribution
$50	24	Final design specification
$0		New capital equipment
$10	2	Initial manufacturing

Joyce has also taken a marketing class, so she examined the possibilities for sales of this product. To begin with, local stores offer specialized manual equipment at a cost of $35 to $40. Although exact estimates are not possible, one store whose stock and level of business seemed "average" admitted to annual sales of 20 compactors. The yellow pages yielded a count of 25 such stores in the local area.

Joyce suggested that from the number of stores selling compactors, the level of sales would increase by 150 per year for the next five years. Later expansion to two nearby metropolitan communities should support similar increases in demand. Mr. Strong asked the team to estimate when the product will break even (ignoring the time value of money).

Options

1. Calculate the equivalent annual worth over a 15 year life considering no inflation and a discount rate of 5%.

2. Instead of assuming that sales increase by 150 per year, reevaluate the project assuming a 20% rate for growth in demand.

Suggestion to the Student

1. The alternative in this case is easy to identify, but there have been at least three objectives identified. What are the two non-economic objectives?

2. You must somehow derive a wholesale price for the can cruncher. Typical markups at the manufacturing and retail levels might be 30% to 40%.

3. What is the fixed cost for continued development?

4. What is the variable cost for production, and how much contribution or profit is there per unit?

5. Calculate the cumulative production, the total cost, and the total revenue over time. What is the breakeven point?

6. Given the economic results and the non-economic criteria, what do you recommend to Mr. Strong?

For The Options:

7. If the growth in sales is projected at a constant amount per year, then an arithmetic gradient can be used for a simple equivalence equation. If the growth is at a constant rate, then a geometric gradient must be used. This is best analyzed with a spreadsheet or an "equivalent" discount rate. Table 5-3 is one way to structure the analysis for comparison with manual calculated values.

Table 5-3		Growth in Sales		
Year	#	Cost or Contribution ($P/F, i, N$)	PW	Cumulative PW
0				
1				
2				

Case 6

Lease a Lot

Round Table Rental Yards provides construction equipment, trailers, crutches, etc., on short-term rentals. Historically, Art, the owner, has purchased the items that he rents out, but his business has been expanding so rapidly that he is considering both straight leases and lease-purchase arrangements. He has decided to use the procurement of a new bulldozer with a list price of $290,000 as a test case.

If he purchases the bulldozer outright, then he must also decide whether he should plan on overhauling it or selling it after 3 years. This overhaul will cost about $150,000, but it should double the useful life of the bulldozer. However, the bulldozer's value on the used market would drop from $180,000 after Year 3 to $135,000 after Year 6. Its annual operation and maintenance costs will start at $25,000 and increase by $7500 each year. This increase is due to increased use more than to increased age, so it is not affected by the overhaul.

The manufacturer has a subsidiary that specializes in financing through leases and lease-purchases. In both cases, the subsidiary uses a term of 5 years with no option to extend it further. Art believes that other contract periods could be negotiated, but for this initial analysis he believes that their standard term is representative of the other possibilities. For the standard lease, the annual payment is $45,000. For the lease-purchase, the annual payment increases by $42,000. Although lease contracts can be written either way, for this lease Art would be responsible for the overhaul cost at Year 3.

Art will insure the bulldozer for theft, catastrophic damage, and liability. This policy will cost him $9500 each year. He will spend about 5% of the rental income transporting it to and from job sites. On the plus side, he expects it to bring in $175,000 the first year. Rental income should increase by $30,000 each year until it hits a maximum utilization of $300,000 per year.

If secured loans are available for 9%, which financing plan do you recommend?

Option

Art's business can depreciate the bulldozer under a 5-year modified accelerated cost recovery system (MACRS depreciation schedule, with a combined state and federal tax rate of 41%. Do tax considerations change your recommendation?

The Board Looks to You

The financial department of Delphi Consolidated Industries (DCI) is just starting a 2-week retreat in the Canadian wilderness. The president of DCI has been approached by an investment banker who has acquired $1,000,000 in DCI bonds in one of his deals. He is offering to sell the bonds back to DCI for $900,000. The president thinks this seems like a good deal for DCI, but his background is in marketing, and he knows his limitations. DCI has had dealing with this investment banker before—his firm often underwrites DCI's bond issues. The president remembers that you seem to know your way around capital investments and has asked you to recommend a course of action by September 13th (later this week). The offer to sell at this price is good through this date.

The bonds consist of 100 bonds, each of which is identical. Each bond was issued 4.5 years ago and had a term of 10 years. The face value of each bond is $10,000, and it pays 14% in quarterly installments. The next payment will be made to the owner of record on September 20th (this is the 18th payment).

When the 100 bonds were sold 4.5 years ago, they brought in $920,000 ($950,000 less $30,000 in selling expenses).

The president and the board of directors have been talking about ways to use the unexpected profits from the sale of some land to the state for a right of way. One idea that was under consideration at the last meeting was to "call" (or pay off early) up to $1,500,000 of an older DCI bond issue.

The older bonds have a face value of $100,000 each and pay 18% in semi-annual installments. They have an early call provision for a 5% premium over face value. The bonds were sold 8 years ago and had a 12-year term. A payment (Number 16) was made last week. These bonds were sold at a premium ($110,000 less $2,500 in selling expenses) in part because of the early call premium.

The pre-tax MARR of DCI is 17.5%.

What do you tell the president to do?

Picking a Price

Sam, the youngest of four, will graduate in industrial engineering this June. His future plans have not solidified yet, but his parents clearly believe that he will be self-supporting. In fact, they are planning on selling their home, taking a world cruise, and investing for their retirement.

Sam has been asked by his parents to spend part of his semester break/holidays helping them analyze a 4-plex that they are considering buying. The building is part of a rental complex with cooperative management of the pool and parking areas. The complex is about 5 years old, and it appears stable and desirable.

Sam realizes that other investments might be more appropriate, but they are not interested in his general advice. Rather, they have asked him to calculate the highest price that they could afford to offer (the asking price is $160,000). Also, they have asked him to ignore the impact of inflation and taxes for this preliminary financial analysis since their future financial position and the future tax laws are both unclear.

They have developed some information, but they suspect some may be missing. Since they are gone for the evening, and Sam wants to ski tomorrow, he plans on "guesstimating" any missing numbers. This will give his parents a preliminary estimate, and it will involve them in the iterations to a sufficiently accurate answer.

Financing for the purchase will come in two pieces. The 20% down payment will be part of the proceeds from the sale of their home, while the other 80% will be financed with a 9%,

20-year mortgage. In discussing this interest rate, his parents also mentioned that their long-term investments in the stock market had averaged an annual rate of return of about 11%.

The annual operating costs for the 4-plex, as reported by the current owner, have been about $350 for water and sewer, $150 for lawn mowing, and a $700 assessment from the cooperative pool/parking authority. The renters pay for their own electricity and natural gas. Property taxes are calculated at 1.8% of the assessed value, and properties are assessed at 100% of market prices with biannual adjustments by the city. The city currently appraises the property at $103,000 for the building and $41,000 for the land. Insurance for fire and liability is 1% of the building's value.

Rents for these and other similar units in the cooperative have been fairly stable at $550/month or $500/month for long-term leases.

Suggestions to the Student

1. Real estate deals usually have substantial transaction costs. Realtor's fees average 6% and are paid by the seller. Loan origination fees, title insurance fees, etc., will often cost the buyer 1.5%. These closing costs must be apportioned between the buyer and the seller—for the property purchase and for its later sale. How much occurs now and how much at the problem's horizon?

2. What horizon should be used? What happens to the property at the problem's horizon?

3. How should the property's value at the horizon be determined? What is it?

4. Are there other costs or possibilities that must be allowed for? What are they?

5. What is the maximum purchased price that can be justified?

6. Which variables represent the bulk of the "risk"? What risks are not addressed?

Case 9

Recycling?

Engineered Products Inc. (EPI) is a conglomerate with both manufacturing and service-based businesses. One of EPI's larger manufacturing plants has been asked to increase its recycling efforts or face a major increase in its disposal fees. EPI prides itself on being a good corporate citizen and has committed to taking any and all feasible actions to reduce the volume and weight of material it sends to the local landfill. The local landfill plans to increase disposal fees by $5 per ton. It is offering to rebate $5 for each ton less than the current 12-month average that the plant sends to the landfill. The plant does not foresee any changes in the current levels of waste generation due to volume or product changes.

Currently, the plant averages sending two containers per day to the landfill. The containers average 10 tons of waste when loaded. The landfill charges $40 per ton to receive the waste. The waste hauler charges $80 per load (one container) to transport the waste. The three waste containers are rented for $5 per container per day.

Currently, cardboard, if collected and bundled for shipment, can be sold for $95 per ton. The plant estimates that it sends 2 tons of cardboard a day to the landfill. To collect the cardboard will require one janitorial associate for 3 hours per day at a cost of $18.50 per hour. The baling equipment will cost $22,000 installed, $45 per week to operate, $4500 per year to maintain, and last 12 years. The equipment will have a salvage value of $5000.

Currently, 25,000 wooden pallets per year are scrapped each year because they are damaged or because they are not of the standard size used by the plant. The plant has budgeted $12,000 for a pallet shredder to chip the pallets as they go into the waste containers to reduce their volume and allow the average weight per container to increase to 11 tons. The shredder has no salvage value at the end of its 6-year useful life. The operating and maintenance cost (not including the operator, a janitorial associate) is $3000 per year. The pallets average 13 pounds each.

A pallet recycler has offered to purchase pallets, which are of certain sizes and in good condition. The pallets that are in these sizes and in acceptable condition amount to half the scrapped pallets. The pallet recycler is offering to pay $1.00 per pallet. To sort the acceptable sizes from the scrap pallets will require three hours of labor per day. This job can also be done by a janitorial associate.

Purchasing has identified a company that will pick up the damaged and unusable pallets and process them into wood chips, which this company then sells. The cost of this service is $1.25 per pallet.

The plant works 5 days a week, 50 weeks a year. The minimum attractive rate of return is 15%.

What do you recommend the plant do?

The Cutting Edge

by
E. R. (Bear) Baker, IV
University of Alaska Anchorage

Elroy had been with Barnes Machine Company a year since finishing a BS in industrial engineering (IE). Barnes had been in business for over 50 years, but the company had only recently moved from Detroit to Gainesville, Georgia. The public reason for the move was the economics of the old facility. Privately, based on comments he had heard, Elroy believed a shift to nonunion labor was a larger motive.

Elroy's boss is the production supervisor, Mr. Hill. Because the plant and the workforce are new, Elroy has been conducting time-and-motion studies to establish new production standards. While these were clearly needed, Elroy was impatient to apply other IE tools he had studied.

One Friday, Mr. Hill asked Elroy to attend a 10 a.m. meeting on Monday. Monday morning, Elroy was surprised to join not only Mr. Hill and John Blackburn, the head of manufacturing engineering, but also Mr. Simkins, the head of marketing and several others from sales and marketing. Most surprising was the attendance of the company's CEO, Mr. Barnes, Jr.

The meeting's purpose was to consider a request for proposal (RFP). As Mr. Simkins quickly pointed out, the request came from one of Barnes's most significant customers. The problem, and the reason for the special meeting, was that a successful bid would exceed current production capabilities. Mr. Simkins, in summarizing, said, "Fortunately Mr. Barnes was farsighted enough to have our new facility built with room for expansion."

Mr. Hill agreed: "I see no reason why we should not bid on this proposal. Of course, as John pointed out, we will need new production capability. While this RFP calls for a five-year delivery plan, the total number of parts has not been specified. Since Simkins believes the data will be available before the final proposal deadline, I suggest that we examine the economics of the various different manufacturing alternatives. To that end, I intend to have Elroy here start that study immediately."

Mr. Barnes ended the meeting with, "I'm sure that not bidding won't hurt our other business with them, but they have been a steady customer since my father started the company and I really would like to help them. Besides, whenever we have added new manufacturing capacity, Simkins has managed to sell it to someone. So whatever you do, Hill, don't let Elroy be too pessimistic. Let's get on with it. I expect a preliminary evaluation in two weeks. By the way, John, don't forget about all that extra equipment we have stored from the old plant. You may find something there that will help keep the cost down."

During the next several days, Elroy met several times with Mr. Hill and John Blackburn. John, who had joined the company after it moved, drove to a warehouse in Atlanta to inspect the stored equipment. In a meeting Wednesday, John said that only a new engine lathe would be required.

Hill said, "If that's all we need to bid this job, Mr. Barnes will be very pleased. After all, what will it cost, 15 or 20 thousand?"

"We can probably find one in that price range, Mr. Hill," John said, "but if we are going to consider this as a long-term investment that Mr. Simkins will market for us, I think we should seriously consider one of the automated systems that have become available in the past few years. Remember, this type of equipment usually lasts a long time. I am sure that it will still be serviceable long after we complete this contract."

"OK, John, your point is well made," Mr. Hill replied. "Elroy see what you can find that will do the job. Check with John on the specs, but take a close look at the economics for us."

During the next few days, Elroy found that there were basically four different possible machine types that would do the job ranging from the traditional manual engine lathe to a computer-controlled lathe with robotic load/unload and tolerance checks. From the manufacturers, he obtained the information contained in Table 10-1.

Table 10-1	Cost Data	
		Annual
Machine Type	Purchase Cost	Maintenance Cost
A. Manual	$18,000	$1,350
B. Semiautomatic	27,000	2,430
C. Automatic	64,000	4,250
D. Automatic with robotic load/unload	120,000	14,400

Machines A and B would each require a full-time operator. A single operator could service two of Machine type C, and Machine type D would require no operator at all. After consulting with John about the skill level required, Elroy checked with accounting and found that an operator would be paid at $14.29 an hour. A 25% incentive is added to base pay for employees on the second or third shifts. In addition, fringe benefits would run 63% of base pay, and manufacturing overhead would be assigned at a rate of 47% of the operator's direct pay. Accounting had indicated that they would try to classify the equipment in the 5-year life category for tax depreciation purposes.

Mr. Hill, John, and Elroy decided that the analysis should be based on production runs of 1000 pieces due to uncertain availability of storage space. Elroy noted that each of the machines has a different production rate and setup procedure. Each manufacturer claims an expected life of about one million pieces. John pointed out that the machines all use the same cutting technique, which implies that the tool and material costs should be about the same. Elroy summarized this in Table 10-2.

Table 10-2		Production Data		
Machine	Setup Cost	Production Rate (Pieces/Hour)	Material + Tool Cost/Piece	
A	$ 750	6 pieces	$0.50	
B	1000	12 pieces	0.50	
C	3000	30 pieces	0.50	
D	6000	30 pieces	0.50	

John pointed out that there is a part currently purchased from an outside vendor that could be produced on this equipment. He estimates the setup cost to be about the same and the production rates to be approximately twice as many pieces per hour for Machine A, about 50% greater for Machine B, and remaining about the same for Machines C and D. Machine tool and material cost would run about $0.70 a unit.

When Elroy checks with accounting, he finds that they purchased about 7,000 of the parts last year. Marketing expects that to increase to 8,000 parts this year and remain steady for a while. Accounting tells him that the average cost per purchased part is $4.26.

In previous economic studies of capital purchases, Elroy has been told to use an interest rate of 15%. He believes that he should do the same here.

Friday afternoon Elroy sits down to begin his analysis. He knows that everyone at the meeting next Monday will expect him to have an answer and that it is very likely that his report will determine whether or not Barnes responds to the RFP.

Case 11

Harbor Delivery Service

Harbor Delivery Service (HDS) is an over the water delivery service operating in several large port/metropolitan areas. Each branch office has from 5 to 15 boats in its fleet. Currently, each branch office purchases its boats locally based on the branch manager's preferences. This has resulted in each branch having a mix of brands and models and both diesel- and gasoline-powered units in some ports. Maintenance for this mixed fleet is a major headache, and costs seem out of control. To better utilize resources, the company has been repositioning boats to avoid unnecessary purchases and idle resources. This has been far from a resounding success, as the receiving locations are not prepared to maintain the boats if they differ from those it currently has. The branch managers inevitably find major faults with the boats transferred into their site. Additionally, this causes the sites to need both diesel and gasoline refueling facilities, with the inevitable confusion and mistakes. The various types and brands also make it difficult to create a "brand image." HDS has decided to centralize procurement of boats and to standardize on brands and fuel types.

The task of standardizing the fleet has been assigned to a team consisting of the chief operating officer and three branch managers. The team has identified the size and configuration of boat that best meets the general needs of HDS but have been unable to agree on a common power unit. A poll of the branch managers finds that five out of ten branch managers prefer the gasoline option due to its higher speed, while two out of ten are indifferent to the choice of power unit.

Marketing has expressed a preference for diesel power units. They claim that the customers perceive diesel units as less flammable and support this preference with data that shows that insurance premiums are $500 more per year for gasoline-powered boats. Marketing cannot show that demand has been impacted by power unit choice.

You have been tasked with recommending the appropriate power unit. To support this task, you have constructed the following table (Table 11-1) based on the specifications of the two boats under consideration.

Table 11-1   Boat Specifications

	Gasoline	Diesel
Purchase price	$76,586	$97,995
Engine size	350 hp	300 hp
Average speed (manufacturer's estimate)		
Knots (nautical mile per hour)	21.1	17.4
Fuel consumption (gallons per hour)	26	17
Fuel capacity (gallons)	300	300

The boat manufacturer (the only difference in the two boats is the engine) has supplied an estimate of the average speed of each unit and the fuel consumption based on this average speed. Since the boats are used in harbors and for fairly short runs, the higher speed of the gasoline engine is valued at only $50 per day. When not in use, the gasoline engines will be turned off, while the diesel units would idle and burn fuel at the rate of 1 gal per hour. Both units are seen as adequate to meet the delivery schedules/requirements of HDS.

Your investigations into maintenance costs have determined that the diesel unit requires $9000 in annual maintenance (mainly for the cooling system), while the gasoline engine unit has an annual cost of $6000. Oil changes are $25 for the gasoline unit and $57 for the diesel unit. Oil changes occur every 100 hours of engine use.

Diesel is estimated to run $2.95 per gallon while gasoline runs $3.15 per gallon. The branch offices are located adjacent to a fueling/service dock ran by another business unit of HDS's parent company. The boats are docked at the fueling facility overnight and each evening the tanks are topped off before the boats are turned over to the maintenance crew for service and cleaning. Thus, nightly refueling stops cost $15, but if refueling must be done during the day it costs $55.

The units will typically cover 200 nautical miles in the course of the day. Crews are changed every six hours. The delivery service operates 18 hours per day 7 days a week.

The diesel units, if purchased, will be kept in service for 4 years before being sold for $48,000 each. The gasoline units will be sold after 3 years of service for $38,000.

HDS's minimum attractive rate of return (MARR) is 18%.

Option

How many nautical miles per day must be traveled to change your recommendation?

Case 12

Buying a Dream

Ms. Sally Firth has worked as a design engineer since graduating from North Central State. She has shifted jobs twice, and she expects to shift again in the near future. All of these jobs are in the same area where she has decided she wants to live for at least the next five years. Thus, she is ready to buy herself a *small* home and move out of her rented apartment.

Ms. Firth has found a home she likes, and she believes that she should buy it before she changes jobs. If she waits to buy, some financial institutions may down rate her due to a short of time on the job in spite of the salary increase she expects. She knows she could get a mortgage, but difficulties in qualifying might restrict her choice of home.

This really concerns her because the home she wants to buy will cost about $96,500 plus closing costs. In fact, the current owner has accepted her offer and given her six weeks to finalize the financing and arrange for closing. He also provided details on the upkeep costs of the house, which she included in her budgeting.

She has set aside about $7500 for a down payment, and she has budgeted for a monthly payment of $900. She expects that her salary will increase about 5% per year in real terms, but she would like to use that increase for "fun" purposes. Until last year, she was still paying off her student loans, and she has not been able to live in the style to which she would like to become accustomed.

She has identified three financing possibilities, but she must compare their effective annual interest rates and other differences to determine which is best.

Current mortgages cover a 30-year period and are available at 10% interest rates with a 5% down payment. With new financing, title insurance is 0.5% of the property's value, and the loan origination fee is 1% of the loan's face value.

However, when the current owner purchased the home, interest rates were only 7%. That mortgage has 25 years to run, with a remaining balance of $68,747.38. The monthly payments also include a reserve allowance for fire and liability insurance and for property taxes. Annually these total $675 for insurance and $850 for taxes.

While the mortgage can be assumed, with the only cost being a $350 charge for credit checks and paperwork, the current owner's equity has to be covered somehow. The bank with the original loan will issue a second mortgage at 12% for a term to match the first mortgage. The requirements for down payments, title insurance, and loan origination fees match those for new first mortgages.

The current owner is also willing to accept a second mortgage directly. While the up-front fees can be omitted, the loan has a 12% interest rate and a 10-year term.

Use annual payments to simplify the analysis.

Options

1. The loan office of the bank has just called Ms. Firth to mention another financing possibility—the graduated payment mortgage. Terms, interest rates, and fees are the same as a normal mortgage. However, by lowering the payments in the early years, it is somewhat easier to qualify for and to afford the home of your dreams. For the first 4 years, the payments are only 80% of the level of a normal mortgage; then for the next 6 years, the payments for both mortgages are the same; and then for the last 20 years, the graduated payment mortgage has higher payments.

2. Although she ignored inflation initially, Ms. Firth recognizes that the rates of the various mortgages include allowances for expected levels of future inflation. This was really made obvious when a friend, George, described his new variable rate mortgage. Everything was similar to what she was used to except that the initial rate (last month) was only 6%. In turn, he agreed that once a year the bank could adjust his rate according to movement of the "prime rate." Thus if inflation is high or money is "tight," his rate could rise as much as 1% per year. There is a lifetime cap of 5% on these increases.

3. Sally is also wondering whether she is better off using some of her "extra" savings to reduce the loan by increasing her down payment, or whether she should use it on early payments.

4. Instead of using the interest rates in the case, research current interest rates and calculate how expensive a home Sally could afford. How do programs such as the Nehemiah Program or state programs for first-time home buyers in your state influence your answer?

Guaranteed Returns

Mr. Juan Tobias Rich (Toby to his friends) has recently inherited $50,000 from the estate of his great-aunt. He has never had capital to invest before, so he is not quite sure what to do. His friend Mr. Richard Stuffy is an investment counselor for a local stock broking firm, Bullfinch and Bearwallow.

In their initial talk, Toby said, "I want to invest the entire inheritance, but no more now. I am living very well on my salary, and I really have no need for the money in the near-term. My goal is to combine this with a good chunk of my future raises, so that I can retire early—in 25 years or so." Richard believes Toby will follow his savings plan. Toby in two years saved enough for the down payment on a house, which he purchased last year.

Toby's capital is too limited for many forms of independent investment. On the other hand, his possibilities go far beyond a certificate of deposit or a few shares of stock. As is normal practice with new clients, Richard has developed three distinctly different investment alternatives (in this case, each is for $50,000). Richard uses his client's reactions to the investment choices to develop clearer goals for each client's investment strategy.

The first of these alternatives is the financing of a second mortgage on a commercial structure. The risk on this investment is relatively low because the loan is secured by the building. It runs for a 25-year period, with annual payments at a 15% annual rate. The borrower pays the cost of title and fire insurance as well as the fees to the bank that acts as the intermediary for payments.

The second alternative is the purchase of 25-year bonds issued by the local power authority. These bonds carry a face rate of 9% with interest paid annually. However, interest rates for this class of bond have risen in recent years (currently 11%) so that the bonds are selling at a discount from their face value. The power authority has recently announced some cost overruns on the nuclear facility that is under construction. In reaction to this news and fearful of more substantial problems, their bonds in particular are being discounted more heavily—to correspond to a 14% rate of return. Thus, if the bonds are paid off, this is a rare opportunity, but there is some risk of default.

The third alternative is as a limited partner in a business run by some friends of Toby and Richard. The business is a spin-off from a local high-tech firm, which is still in the embryonic stage. Their friends hope to take the firm public (sell stock and become publicly held) within three years. At this point, the investor's capital would be returned with substantial interest. However, the company may go bankrupt instead. If this happens, Toby can expect to get back about 20 cents on the dollar. Although many intermediate states are possible, Richard believes that the two extreme possibilities provide adequate guidance. He estimates that there is a 40% chance of success and that the original investment will increase tenfold if the investment is a success.

Toby's (8%) mortgage is his only outstanding debt. Therefore, a fourth alternative would be to pay $50,000 of it off early. Each of these investments has a higher rate of return, but he is not certain how he should pick one. The question is further complicated by Richard's comment that the current capital shortage has pushed interest rates and other returns up from the normal rate of 10% for relatively risk-free investments.

What do you recommend that Toby do? Why?

Options

1. After taxes: If Tobias is in a 35% bracket for income taxes (state and federal combined), and if only 40% of a capital gain is likely to be taxable at retirement, then which investment is better?

2. If inflation seems likely to rise to about 5% from its current insignificant level, then which investment seems better?

3. Describe the relationship of risk to return for the three investments. Are all three on the "efficient frontier" for the risk/return trade-off?

Suggestions to the Student

1. The first two alternatives clearly have a 25-year life, which matches Toby's time horizon. How should the more risky investment in his friends' venture be evaluated: over 3 years or over 25?

2. What should be averaged to measure the value of the third alternative—the dollar returns at year 3 (or at year x), the present worth, the future worth, or the internal rates of return?

3. What is the best criterion: maximizing present or future worth, choosing the highest internal rate of return, or some risk adjustment of these? What is the best discount rate to use?

4. What is your reinvestment assumption, and is it the same for each alternative?

Case 14

Northern Gushers

Northern Gushers Drilling has developed a lease on the North Slope of Alaska over the last five years. They have drilled 16 production wells evenly spaced over the four square miles of the lease tract. Every well's production declines over time, so to maintain a "steady" total flow new wells are drilled. Specifically, total production from the existing wells will decline at 17% per year if no new wells are drilled. All wells have been directionally drilled from the gravel drill pad, which also contains a processing facility (see Figure 14-1).

Figure 14-1   Field Arrangement

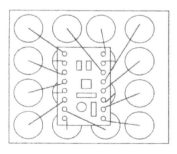

This processing facility separates water and natural gas from the crude oil stream. By reducing the pressure from formation to atmospheric levels, the volatile gases are removed from the oil. The oil is then dehydrated to remove water before transfer to the pipeline. As shown in the flowchart of Figure 14-2, a small portion of the natural gas is used to power the facility. Then most of the natural gas and all the water are repressurized and reinjected into the formation. This reinjection avoids the environmental problems of flaring the gas or surface disposal of contaminated water. It also helps maintain the pressure in the oil formation to increase total recovery.

Figure 14-2   Processing Flow Chart

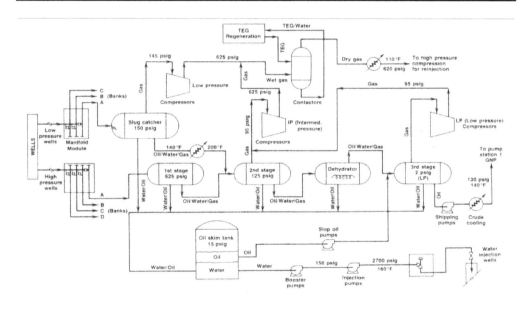

The timing of new wells has been planned to maintain a steady flow of about 20,000 barrels of oil per day (BOPD). This flow rate is the "shipping space" on the Great Northern Pipeline that has been allotted to Northern Gushers parent company. Fluctuations in production are matched with the shipping space by buying, selling, and trading with other shippers at about the tariff's cost per barrel. A second consideration has been maximizing total recovery by distributing new wells over the leased tract. Thus, each of the 16 wells has helped maintain current production and also increased total recovery from the field.

Now Northern Gusher is facing a different problem. The entire leased tract has been covered by the 16 wells. New wells will be drilled "in-between" existing wells and will therefore have less impact on total recovery from the field. Each new well will increase production now by 2000 BOPD and increase the decline rate by 1%. Since each well costs about $2 million to drill and $1.75 million to tie into the production facilities, the management of Northern Gushers must justify this decision to their parent company by identifying the rate of return on the required capital investment. Additional capital is required every 7 years to do a well work-over for $1.25 million. Abandonment costs in the final year of production amount to 10% of the initial drilling and facility costs. The abandonment costs are required by state agencies to return the land to its initial condition.

Currently, the tariff for transportation through the pipeline is $5.25 per barrel, and another $3.75 per barrel is required to ship it to market. The incremental annual operating and maintenance cost for the field is $200,000 for each new well.

At this point, Northern Gusher must decide whether to initiate planning and construction for Well 17. This particular well could come on line next year with an estimated production rate of 2000 BOPD once tied into existing separation facilities. By next year, Northern Gusher's total production rate will fall to 18,000 BOPD. The production of the 16 wells is declining at 17% per year. With the new well added in, the higher production rate results in a field decline rate that is 1% higher at 18% per year.

For simplification, the new decline rate can be assumed to begin as soon as the new well is drilled since the field will be producing at a higher rate almost from the start. While more wells will be drilled in the future, economic analysis of Well 17 is done without considering them. Oil production at the facility will be closed down (the field will be shut-in) when the total field production reaches 500 BOPD. At that production rate, it is no longer economic to operate the field.

Because some of the oil produced by Well 17 would have been produced in later years by nearby wells, the incremental "production rate" with Well 17 versus without Well 17 will be negative in later years. This sign change in the cash flows can result in multiple rates of return, so that is an additional concern of management.

The value of oil has varied dramatically over the last six years, from a low of $18 per barrel to a high of $140 per barrel, back down to $22 per barrel and then back up to the current level of $45 per barrel. Because of this vast uncertainty, your boss has given you guidelines of $30 per barrel and a horizon of 20 years for the initial analysis.

Should Well 17 be added now, and how much additional oil will be produced? What is the incremental rate of return on this investment?

Options

1. If management demands an internal rate of return of 15% on investments, can Well 17 be justified now or at a later date? If now, when can well 18, 19, and so on be justified? Each of these later wells will have a similar effect on the total decline rate for the field. Specifically, assume that each well increases the decline rate by 1%.

2. Rather than considering the price of oil to be "fixed" at its current level, consider the impact of a higher or lower inflation rate for oil than for the economy as a whole.

Case 15

Pave the Stockpile Area?

by
Herb Schroeder
University of Alaska Anchorage

A contractor operates a rock crusher and stores the material until needed in an adjacent stockpile area. The stockpile area is unpaved, and some of the produced material is contaminated by the sub-grade and cannot be used. The estimated stockpile losses are shown in Table 15-1. This equates to a combined loss of 8% each year.

Table 15-1	Current Stockpile Losses

Material Size	Loss
¼" minus	12%–20%
¾" minus & ¾" x # 4	8%–15%
1 ½" minus & 1 ½" x ¾"	4%–10%

The contractor is considering paving the stockpile area to reduce the loss of material. Paving the stockpile area would reduce the losses due to contamination to about 2%. The crusher produces 250,000 tons/year at a cost of $1.50/ton. The new paving should last 10 years, and there is no salvage value.

The paving will cover 4.56 acres. The previous stockpile losses will serve as sub-grade and base course for the new surfacing. A 3" thick surface will require 3700 tons of asphaltic concrete, which costs $20/ton installed since he can supply the material at his cost. The engineering and site work involve a one-time cost of about $5000. The surface requires routine maintenance costing about $1000/year.

The contractor has estimated lower and upper limits for the data as shown in Table 15-2.

Table 15-2	Lower and Upper Limits on Estimated Data	
Economic life	−50%	+100%
Price/ton	−20%	+40%
Tons/year	−40%	+20%
Loss with paving	−20%	+20%
Design cost for paving	−10%	+10%
Maintenance	−10%	+20%
First cost for paving	−5%	+10%

1. Determine the rate of return on the repaving, and recommend whether the contractor should pave the area?

2. Use breakeven charts or a spider plot to analyze which uncertainties could change your recommendation.

3. Construct a tornado diagram to summarize your results for management.

Case 16

Great White Hall

Flatland Views has advertised for proposals to build a new community center, but the city council cannot agree on how to evaluate the submitted proposals. The request for proposal (RFP) specified that respondents had to meet certain basic needs, although optional items could be included. The RFP also asked that each respondent calculate a benefit/cost (B/C) ratio using a discount rate of 12%. The RFP did specify the approximate size of each optional facility and the use that could be expected (and the value of such use in dollars per hour).

The RFP stipulated that the council would select a package of facilities based on estimated construction costs and B/C ratios. Since this package might not match any proposal, the council could issue a new RFP. However, if a new RFP is issued, only respondents to the first RFP may respond. The council's intent is to provide an incentive for participation in the first RFP. Instead of a second RFP, the council could choose to simply negotiate with one of the original bidders.

Three firms responded to the RFP, but they used different assumptions on how to calculate the ratio as well as including different options within their proposals. Their construction materials and associated lives are similar, but their designs differ substantially. The proposals can be summarized as follows.

Tightfisted Proposal

Averell Johnson, the conservative patriarch of the city's construction community, has proposed a bare-bones facility (see Table 16-1). Assuming 50 years of use and end-of-period cash flows, his proposal has a B/C ratio of __.[1] His proposal also assumes that construction expenditures are all made at the start of the construction period.

Table 16-1	Tightfisted Proposal
Construction:	1 year: $2.5 million
Annual operation:	Gym: $120,000
	City offices: $190,000
Annual benefit:	Gym: 60 hours/week at $200/hour

Major Projects Proposal

The proposal that has been supported by the "town and gown" crowd includes a small auditorium/theater and a library as well as the gym (see Table 16-2). Major Projects Inc. has evaluated the proposal over 30 years of use for the benefits and for a 12-month term for the construction phase. Major Projects has assumed end-of-period cash flows, but they have analyzed the construction phase as 12 months—each with an equal share of the construction expenditures. Their calculated B/C ratio is ___.[1]

Table 16-2	Major Projects Proposal
Construction:	12 months: $4.8 million
Annual operation:	Gym: $110,000
	City offices: $165,000
	Library: $450,000 (mostly salaries)
	Theater: $65,000
Annual benefit:	Gym: 60 hours/week at $200/hour
	Library: $0.5 million in improved education
	Theater: 16 hours/week at $450/hour

---

[1] The omitted B/C ratios for each facility are not necessary for the rest of the case. The "easiest" option is to calculate them.

Energy Breakthrough Proposal

The third proposal (Table 16-3) is from a new firm that specializes in the design and construction of energy-efficient structures. They based their B/C ratio, _____, on assumptions of 40 years of use and costs and benefits that flow continuously over that time (distributed rather than end-of-period cash flows).

Table 16-3	Energy Breakthrough Proposal
Construction:	1 year: $3.9 million
Annual operation:	Gym: $ 65,000
	City offices: $100,000
	Theater: $15,000
Annual benefit:	Gym: 60 hours/week at $200/hour
	Theater: 16 hours/week at $450/hour

The Council's Solution

Overwhelmed by the responses, the city council has decided to hire you as a consultant. Your contract requires you to calculate comparable ratios, to recommend a package of facilities, and to recommend a contractor.

Options

1. The problem can be simplified by specifying that all projects assume end-of-period cash flows except for construction costs, which could be specified to occur before construction begins. *This may or may not be the best assumption.*

2. The problem can be simplified by limiting it to the calculation of the omitted B/C ratios.

3. The problem can be simplified by reducing the scope of the consultant's contract to constructing valid comparisons of the three proposals.

A Free Lunch?

by
William Truran
Stevens Institute of Technology

Zippy Quick, in his bright suit and straw hat, walked briskly into the large building complex to see the superintendent (Super) about her heating, ventilating, and air conditioning (HVAC) equipment. Zippy said "How'd you like to hear my scheme for you to make money from nothing and then end up with some new equipment that you can effectively use?"

Well, the Super was interested but suspicious. "What do you mean, I have to put up no money and end up with new equipment I can keep?"

"Correct, I'd have the specialists come in here and do the installation and startup of the equipment. You pay nothing, AND, you'd be *making* money in a year."

"Where's the catch?" exclaimed the Super.

"NONE, no catch" said Zippy. "Just sign a three-year contract to split 50/50 any savings the equipment generates on the use of your HVAC equipment plus pay for a service contract for the equipment's maintenance. We will take out the loan for the equipment and installation and pay off the loan with our share of the savings. After three years the equipment is yours. You'll only be giving us some of the money you are paying the power company for electricity. The service contract for the equipment is $15,000 per year, but think of your peace of mind!"

The Super is suspicious of Zippy, but he seems to know HVAC equipment. The deal he is offering seems to be risk free if she gets a clause in the contract that gets the equipment removed at no cost after the first year if she is not seeing cash flow in excess of $15,000 to cover the maintenance contract.

Based on her own research in response to Zippy's claims and some questions she asks him, she finds the following details to use in trying to evaluate Zippy's offer:

- The combined horsepower (HP) for the fans and blowers in the HVAC system total 400 HP. The system currently runs at capacity 100% of the time.
- Most systems are overdesigned. A 10% reduction from maximum speeds is possible without impacting performance if fixed speed motors and controls were originally used to drive the blowers. This system was designed that way.
- Zippy claims that with his equipment (variable speed drives and controls for the motors) the system will run at 20% of capacity at least 45% of the time. For example, on nights, weekends, and holidays, there is little need for heating and cooling.
- While not part of Zippy's pitch, running the fans and blowers at less than capacity will extend their life from about 7 years to 10 years, which is also the life for Zippy's system.
- Electricity is currently costing $.12 per kilowatt hour.
- All the fans in the HVAC system are centrifugal—thus the effective HP is a cube of the speed (see Figure 17-1). This implies at 90% speed, the required HP is $.9^3$ times the required horsepower at 100% of capacity. Note that at 20% speed, the required HP is actually 10 HP rather than the 3.2 HP predicted by $.2^3$ times 400 HP.
- One horsepower requires 748 watts.

Questions to answer:

1. Is Zippy off of his rocker? (HINT: Yes, quite a strange way to dress and present himself; but beside his weirdness, he may have something to provide him money to buy a new suit.) Is this a good deal for Zippy? Zippy has a minimum attractive rate of return (MARR) of 12% and does not really take out a loan.
2. Considered over three years is this a good deal for the Super? Is this a free lunch? The Super has an MARR of 9%.
3. What should the Super do?

Options

1. What is the PW at the Super's MARR and the rate of return for the system considered over its 10-year life?
2. For Zippy's original proposal, what is the payback period for Zippy? For the Super?

Figure 17-1   Speed vs. Horsepower

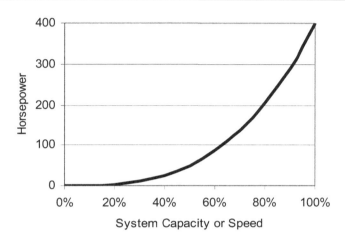

Case 18
Gravity-Free High

Inspired by late night conversations, four friends have decided to form a high-tech start-up company. Glenn works in the space industry. The other three work for a pharmaceutical company. They intend to produce a variant of mood-altering drug used in psychiatry. The production process is slow, and it requires both heating and a very precise mix of ingredients. When produced on earth, the heating produces density changes. The resulting convection currents change the local mixture, and then the process may halt with a contaminated product. If the process is conducted in a "weightless" environment, then the density changes do not induce convection currents, and a purer drug is obtained.

The four friends represent an auspicious mix of experience, enthusiasm, energy, and attitudes. As the career and expertise of each are based directly on their academic training, that may be the easiest way to introduce them. Glenn majored in aeronautical engineering. Joe received an MBA with an emphasis in marketing, while Francine concentrated on finance for her MBA. Fred received his doctorate in biochemistry.

Glenn no longer does engineering design; but instead, he manages projects. He readily identifies where things go wrong, and he has a broad understanding of the cost end of the launch business. He may not be a creative genius, but he has very strong analytical skills.

Joe is an idea man whose dream is to be the perceptive leader and conceptual genius for a new product breakthrough. Although he is only 31, his rapid rise through the normally staid ranks of a safety conscious pharmaceutical firm substantiates the possibility of his dream.

Francine is the most senior of all, but she has been frustrated by the persistence of the "old-boy" network in the financial end of the pharmaceutical industry. She is tired of having to be significantly better than male coworkers to be the one to receive a promotion. She anticipates no problem in arranging financing for the firm.

Finally, Fred is a recent graduate who has spent three years on a postdoctoral research assignment followed by only a year in the company's lab. He wants a much faster route to the top than he sees in the bureaucratic maze of a large lab.

For the past year, they have spent weekends and evenings generating, refining, and analyzing ideas. Now they need to obtain financing so that they can go forward. They plan on asking a trio of Atlanta business executives for the initial venture capital. Later they will negotiate the terms, but first they must establish their plan as a good fiscal risk. The four entrepreneurs agree that their presentation must explicitly treat the uncertainty in outcomes.

Here the agreement ends. Francine and Joe are used to scenarios and spreadsheets, while Glenn and Fred are more used to formulas, statistics, and graphs. A portion of their discussion follows.

Joe: "Look, these executives have backgrounds similar to mine and Francine's, and they will be most comfortable with a low, a medium, and a high scenario."

Glenn: "While the three bottom lines are useful, they know the worst case is going belly-up. And that it may be the most likely! The best case involves going public with incredible returns, while in-between is a small loss or profit. We can develop and show them a better understanding through charts of relative sensitivity and breakeven analysis. This way we and they can tell which financial elements are most critical to our success."

Fred: "Why not quantify the probabilities of different profits? This includes the information of the scenarios and adds more. We could state the probability of losing the initial investment, of staying even, or of doubling or quadrupling or even increasing by a full order of magnitude."

Francine: "I believe we must do all of this, but some of it may not be in our proposal. They expect us to be honest, but they also expect that we will show our proposal in a favorable way. I think we should each analyze the problem individually, so that we will have four draft proposals with appendixes showing our complete calculations. Then we can combine them for the best possible proposal."

The four agreed on Francine's suggestion; and before they broke up for the night, they defined a common data set (Tables 18-1 to -4).

Project Start-up

This phase is expected to take six months to arrange financing, locate a facility, order lab equipment, etc. Note that the four have agreed to fix their salaries at their current levels even though their workweeks are likely to double for the next three to five years.

Table 18-1   Project Start-up Phase

Salaries:	$205,000 + 21% for benefits
Legal fees:	$50,000 (primarily incorporation expenses)
Insurance:	$5000/employee-year (unemployment and liability)
Deposits:	$20,000 (75% for lease, 25% for equipment orders)

Production Prototype

This phase lasts 18 months. The laboratory development of the production equipment to be used in space requires 20 lab employees. The secretary hired during start-up will be supplemented with three more office employees. These employees will all be carried over into later stages, and the laboratory and its equipment will be a permanent facet of the company. The company will also be planning and contracting for the later stages.

Table 18-2   Production Prototype Phase

Salaries:	$750,000 additional per year
Legal fees:	$40,000 (primarily for contracting)
Equipment cost:	$2.3 million (half replaced every five years, with another $0.8 million in new purchases each year)
Lease costs:	$180,000 annually

Initial Production

This phase takes a year and adds another 15 employees. The main costs are for the space-based equipment, for launch fees, and for contractual fees of the agency employees who monitor the equipment and recover the product. Because of initial debugging requirements, the costs for launches and agency employees are double what should follow in later years. Similarly, the equipment costs are 50% higher.

Table 18-3	Initial Production Phase
Salaries:	$500,000 additional
Legal fees:	$80,000
Launch fees:	$6 million
Agency fees:	$2 million
Equipment costs:	$3 million (production facilities)

Although this equipment will last 5 to 10 years, the plan is to replace the space-based portion each year. This keeps up with technological progress and the growth in demand. Its chief salvage value is in cutting costs for earth-based production facilities to a minimum—only 10% of what their expected level would otherwise have been. The initial production level is likely to be low, but it should be adequate to cover the cost of the marketing efforts that will begin this year.

Growth Phase

This phase is likely to last 10 years. Although explosive growth for the product is expected, this should not result in any significant increase in the number of employees after the initial hire to support this phase. This initial hire will be made up mostly of field representatives for the marketing effort, along with the office-based support staff, for a total of 20 new employees. This phase will require more liability insurance costing about $3 million per year.

The explosive growth in demand will not increase launch fees, agency employee fees, etc., as most production will be mechanized, and technological advances are expected to counteract increases in size and weight. From an initial level of 200,000 units, demand is expected to grow along the S-curve detailed in Table 18-4. Raw material costs, energy costs, and other variable production costs will be added onto the cost of the product later.

As this is a prescription item, it is not particularly price sensitive. Politically (since National Aeronautics and Space Administration (NASA) is involved in the launches), the price should be set so that the expected rate of return is not too high. The four agreed to use .6, .3, and .1 respectively for the probabilities of failure, OK, and success. They also believe NASA will accept a 25% expected rate of return for the first 10 years of project life (7 years of production).

Table 18-4	Volume (in millions)									
Year	1	2	3	4	5	6	7	8	9	10
Volume	0.2	0.4	0.8	1.6	3.2	4.8	5.8	6.6	7.2	7.6

Options

1. Modify the failure assumption with these conditional probabilities: 20% at the end of start-up, 25% after the prototype, and 30% after initial production. The rest of the 60% failure probability comes after the first year of growth.

2. Add to Option 1 the following. During the growth phase, there is about a 3% chance each year of a competitor developing a ground based and much cheaper alternative.

3. After five years of large-scale use, the Federal Drug Administration may allow the sale of a milder non-prescriptive variant. This euphoric would follow a similar growth pattern, but its maximum volume might be 700 million units in the U.S. alone. The price would have to be cut to $3.50 per unit to compete with alcohol. The production and marketing costs would increase by an order of magnitude, while the legal costs would increase a hundredfold. A key question is whether patent protection can combine with limited access to space to protect the exclusive position of Gravity-Free High.

Crummy Casting

Net Shape Casting, Inc. (NSCI) is having problems with a titanium alloy part cast for an Air Force contract. The part has an extremely complex shape. Nearly 70% of the castings are produced with unacceptable voids. NSCI "solves" this by producing many extra castings and recycling defective ones into scrap/raw material.

This solution more than triples the cost of labor and energy during casting. It also contaminates the alloy during the pouring and remelt stages. Problems with voids are obvious before the expensive machining stages begin, or annual costs would be much higher than the $36,000 extra it now costs. In the long run (an extra 3-year delay), the problem can be solved in the next redesign cycle. The problem stems from the most recent design revision, which was just instituted.

The foundry's floor supervisor believes that better understanding of the flow patterns in the mold and some mold modifications may solve the problem. He asked engineering to estimate the cost of a "research" study. They propose spending $50,000 for lab work, which they estimate has a 75% chance of leading to a workable solution. They believe spending an additional $10,000 for an outside consultant specializing in flow simulation would increase the probability of success by 10%.

Note that the contract requires a fixed number of acceptable parts per year in which the number made (M) times the success rate equals the number needed (N). With the current success rate, $M = N/(1 - 0.7)$. The extra cost per year for the unacceptable voids is proportional to the number of defectives $= M - N$.

If NISC uses a discount rate of 10%, should the research study be conducted and should the consultant be hired?

Options

1. Analyze the sensitivity of your recommendation to the estimated probabilities of success.

2. The basic problem has modeled the outcomes of the two study versions as complete success or complete failure. Instead assume that the internal team with or without the consultant would improve the current 30% success rate as shown in Table 19-1.

Table 19-1	Success Rate After Research Study			
Probability	.1	.2	.5	.2
Outcome	Terrible	Pessimistic	Expected	Optimistic
With consultant	50%	70%	85%	100%
Without consultant	40%	60%	75%	100%

3. The internal study could be started immediately, and then the decision on the consultant could be made in three months. If this option is selected, the internal study will reliably indicate the outcome. If the consultant were hired, it would add two months of delay, add $7000 in internal lab costs, and would halve the consultant's fee. When should the consultant be hired, and what is the expected value of the delay?

Suggestions for the Student

1. The first step is to find the expected value of the three alternatives: two versions of the study and continuing as is.

2. By equating the Step 1's expected-value equations, breakeven probabilities can be found.

New World Mining

The World Bank is evaluating a proposed hard rock mining venture in a depressed region of a developing country. The development would provide badly needed jobs and earn scarce hard currency. Thus, the country wants the project very badly. World Bank financing may be the only source since the country has already incurred substantial debts. The Bank's board has decreed that financing will only be provided if the project is self-supporting, with no subsidy from the country's hard-pressed economy.

The country's currency is unstable, and the mineral is priced in dollars on the world market, so the entire project is being analyzed in U.S. dollars. The discussion that follows summarizes the data collection phase. Your assignment is to recommend whether or not the project should be funded.

World Bank financing, if offered, will be at 8.5% with a level repayment schedule. No payments will be required until the end of the sixth year, and then they will continue for 30 years. As production will start earlier, the World Bank is requiring that 60% of the "initial profits" be deposited in reserve accounts. When prices are high, more deposits will be made into these accounts. Then when mineral prices drop, these accounts will be used to "make" the payments.

The deposit is a mixture of hard rock minerals, but the most important one is used for specialty steels. When demand is high, prices rise; conversely, when demand is low, prices fall. These swings cause variations of over 50% from the average price. Because this project will support a number of new communities, only minor adjustments of its production level are possible when prices fall. This deposit is not large enough to change the worldwide price/demand relationship. For the initial analysis assume that prices and production are stable at their average level.

The country's economists have agreed with the suggestion of using a stable long-term average price, but they are recommending that this include an allowance for long-term increases. They are suggesting that the annual rate be 3%. The data was gathered during a 15-year period when the overall inflation rate for U.S. dollars was 3.5%. The current inflation rate is relatively close to this long-term inflation rate.

The financing for the project will be released in three segments. The first is expected to cover geological sampling, the detailed design effort, and construction of initial camp facilities. The second segment covers the purchase of equipment and the bulk of the construction effort. The third segment includes $15 million for working capital and funding for the last phase of construction.

First-Segment Activities

Exploratory sampling provided rough estimates for the quality and size of the deposit. The detailed development plan requires more sampling. This can proceed in conjunction with development of the basic infrastructure for the site (engineering design is already done). The major items with their associated costs (in millions, as are all stated costs) are summarized in Table 20-1. These expenses and activities will span a year after the go decision.

Table 20-1	First Segment Costs

Cost ($M)	Item
$6.5	Initial harbor and road development
$1.4	Geological sampling
$2.1	Camp development
$1.8	Engineering and planning for mine development

Second-Segment Activities

Ordering and construction of the bulk-handling equipment will take two years. This includes more infrastructure—roads, harbor, and supporting communities. These costs are summarized in Table 20-2 along with replacement costs and intervals. These initial construction costs, like the other construction costs, are estimated in year-of-expenditure dollars. The replacement costs will inflate at about the general 3.5% rate.

Table 20-2   Second–Segment Costs

Cost ($M)	Item
$35	Bulk-handling conveyors, etc.
$18	Every 5 years for replacement
$55	Infrastructure development
$16	Every 10 years for replacement

Third-Segment Activities

There is $10 million worth of construction to be completed during the first year. Production will be at half of the full rate, and it will cost 25% more than "normal." This construction will add more bulk-handling equipment with similar maintenance and replacement needs. At full production, which starts in the second year, the deposits will last nearly 50 years.

Energy costs are expected to keep pace with inflation. Labor costs are expected to grow about 1% faster as living conditions and expectations are raised. At full production with current prices, annual sales of $165 million are expected initially. Past experience with other mines suggests that this will fall about 1% annually because the richest areas of the deposit are developed first. The annual costs for mine operation and infrastructure are shown in Table 20-3.

Table 20-3	Annual Costs for Mine and Infrastructure	

Cost ($M)	Item
$12	Energy costs
$49	Labor costs

Options

1. How far off does each estimate have to be to change the recommendation? For example, the World Bank expects that there will be substantial pressure to increase the pay rates more rapidly than the 1% differential over inflation that was analyzed.

2. Use simulation to study the impact of price fluctuations. Is the initial reserve fund adequate, and what guidelines do you recommend for making deposits and withdrawals from the reserve fund? Note that initial expenditures on the mine will be made only at a time when prices are reasonably good and likely to remain so.

Glowing in the Dark

POC, Inc., treats hazardous wastes at over a hundred facilities in the United States. Their company name comes from an early advertising theme, Protecting Our Children. POC may construct a new facility for a substance recently reclassified by the EPA as hazardous. The unit cost of treatment is low relative to the value of the process that produces the hazardous chemical; thus, the market appears promising.

POC builds small, specialized facilities, with only a few processes used or waste streams treated. This minimizes potential interactions between incoming wastes, transportation dangers, and political difficulties in the approval of new facilities.

The reclassified chemical is a by-product of an older inspection process that relies on visual inspection of a dipped part under "black lights." It is commonly used in foundries and assembly plants for engine parts. Sites where this process is still used are scattered throughout the country, but the biggest concentrations are around Detroit, Michigan and Gary, Indiana. POC plans to build the facility between these two cities.

The EPA has defined new exposure standards, but it has not yet set a timetable for compliance with the permanent standard. It has established a minimum interim standard. Preliminary indications are that the volume of waste will double when the new standard is imposed. Even the interim standard, at pricing levels used in an earlier feasibility study, will produce a net annual revenue of about $900,000 to POC. This should also double when the new standard is implemented.

The timing of the permanent standard depends on an EPA study of economic consequences that will take almost two years to complete. Dr. Eric Klossen, the company's director for governmental relations, guesstimates that the permanent standard will become effective in 3, 5, or 10 years with respective probabilities of 20%, 50%, and 30%.

The director of engineering, George C. Perriwinkle, has had his staff estimate three alternatives. The structure can be sized to meet the volume of the interim or of the permanent standard. If the smaller size is built, it can include some utilities and facilities to support the later expansion. In each case initial construction includes one "treatment process" line and the building to contain it. The differences between the options focus on whether the building has room for the larger "treatment process" line. In each case the equipment of the larger treatment line will only be purchased and installed when it is needed.

Alternative M (Minimal)

The **minimal** facility sized for the interim standard will cost $6 million. Expanding it for the permanent standard will cost another $5 million later. This facility will cost $200,000 annually to operate initially, then it will double when expanded.

Alternative S (Staged)

Construction can be **staged** by sizing utilities, loading docks, etc., to support later expansion. This adds very little to the annual operating costs, but it does increase initial construction costs by $1.15 million. In return, the expansion will only cost $3 million later.

Alternative A (All)

Construction of **all** of the project can be undertaken immediately. Initial construction costs are $9 million, and annual operating expenses increase. However, POC can use the extra building space for warehousing. The value of this use about equals the increased operating cost.

Each construction stage will take about a year, with the bulk of the costs occurring at the start of the year. Other costs and revenues can be evaluated as end-of-year cash flows. Phase-in periods for the revenues to reach their projected levels are short enough that they can be ignored for this analysis. If POC uses a 30-year horizon, and a 10% discount rate, evaluate these alternatives. This evaluation should consider both the expected value and the variance of the return with each option. Which alternative should be selected and why?

114

Options

1. Simplify the case by ignoring the variance of the return with each alternative.

2. POC can wait until the research study is done to begin construction of the treatment plant. Dr. Klossen believes that predictions can be far more accurate then, but still not perfect. If a low economic cost is predicted, then there is a 50% chance each for 1- or 3-year additional delays. If a medium economic cost is predicted, then there is one chance in six of an additional 1-year delay, a 50% chance of an additional 3-year delay, and one chance in three of an additional 8-year delay. If a high economic cost is predicted, then there is a 50% chance each for additional 3- and 8-year delays. According to Dr. Klossen's Bayes' theorem calculations, these are consistent with a 30% probability of a low economic cost prediction, 30% for medium, and 40% for high.

   Should they wait? What is the expected value of the sample information received by waiting until the end of the research study? What is the cost of the delay?

3. If POC starts construction in the near future, there is a 5% chance that a new competitor might start a plant as well. This would cut POC's market in half. If POC delays for two years, this probability goes up to 15%. If POC builds a plant first, it does not expect a competitor to enter this market. What should POC do?

Case 22

City Car

National Motors of America has been impressed and distressed by the initial sales of small, very low-cost imports from Eastern Europe, Malaysia, and Korea. These imports are "bare-bones" autos, but at prices of $4000 to $5000 they are very popular. With the much higher labor rates of U.S. auto plants, National cannot compete directly with these autos. National could contract with the foreign automakers and simply distribute imports through its dealership chain—with or without a National nameplate and logo.

Mr. Joe Mercer, the president of National, wants to emphasize a leadership strategy instead. He believes (and his belief is substantiated by market research) that many of these cars are destined for short-range in-town trips with only one or two passengers. He believes it is now time to dust off the futuristic tricycle cars that were suggested during an energy crisis. Their simplicity, small size, and low material cost should allow National to produce them at a cost competitive with the new bare-bones imports. Furthermore, their uniqueness and aerodynamic sophistication will make them more difficult to produce in low-technology, low-wage countries.

When pressed by one of his many doubters, Mr. Mercer stresses the necessity of selling vehicles for each part of the market—including the entry level. In total, the U.S. market for cars with "entry-level amenities" is about one million autos per year.

He also identifies the key strategic disadvantages of the two other competitive responses. First, the imports are very similar to the domestic models in the countries they come from, and they are very dissimilar from the rest of National's line. Thus the offshore manufacturers have greater economies of scale in production. Second, putting their label on an import can be easily copied by a U.S. major competitor. In fact, National and all of its competitors are already doing this. This provides cars to sell at competitive prices, and it maintains complete lines of products (essential for brand loyalty), but it does not provide a strategic advantage for National.

This could be another Edsel, but Mr. Mercer has decided that the potential gains in sales and in company and personal reputation are worth the risk. National can move rapidly with this innovative design, because the bulk of the design effort was completed before the designs were shelved. Rapidly means that in two years the first car can roll off a modified assembly line. Thus, revenue really begins with the third year after the decision to proceed.

Mr. Mercer does not really expect the first generation of tricycle cars, the TRIO, to provide large profits to National. He does expect this head start to translate into an enduring large share of a new, growing market. Thus, he plans to convert the Joliet plant (which is due for renovation) to the new car, and then, when a second generation is due (about five years later), he plans to build at least one new dedicated facility. Each facility can operate at 50,000 to 200,000 vehicles per year, depending on the number of shifts.

He has asked you to include the impact of this second-generation decision in your analysis. He envisions three options: first, a high demand coupled with at least one new dedicated facility. Second, the TRIO might be moderately successful, but not able to justify construction of a new facility. Third, the TRIO might be canceled entirely after the first generation. Then the Joliet plant would be converted back to production of standard compacts.

The standard interest rate that National uses is 8%, along with a 30-year life span for most plants and 10 years for most production equipment. Mr. Mercer has acceded to the requests of his critics and limited the analysis of the TRIO to the first and second generations. The third generation is more speculative, and it is conservative not to include it.

National's subcompact and compact plants are running at only 65% of capacity, thus Joliet's current output can be absorbed by the other plants. The conversion process will cost $75 million more than the renovation that was scheduled for Joliet. The extra equipment that is replaced will have a salvage value of $14 million. A new plant would cost $450 million, with 60% of this being for equipment.

Mr. Mercer had the marketing department develop tables that summarized their expectations for the TRIO's market (Tables 22-1 and 22-2).

| Table 22-1 | **Sales for First Five Years** | | | |

			Annual Sales	
Level	Probability		First Year	Last Year
Low	.40		30,000	50,000
Medium	.35		50,000	100,000
High	.25		75,000	200,000

| Table 22-2 | **Sales for Second Generation** | | |

	Probabilities for 2nd Generation Sales Given 1st		
First-Generation Sales	Second-Generation Sales		
	Low	Medium	High
Low	.6	.3	.1
Medium	.3	.5	.2
High	.1	.4	.5

For the first year National expects to sell the cars for about $1350 more than their variable costs for production, marketing, and transportation. In later years, this margin should improve by $50 per year as production becomes more efficient. There will also be a $100 drop in variable costs when (and if) a new facility, designed specifically for this product, is built. After five years at Joliet and two years at the new facility, the rate at which costs will drop slows to $25 per year. By this point, costs will have been reduced substantially and further savings become more and more difficult.

Mr. Mercer's assignment to you follows. At these markups, are there reasonable trends in market demand for the second generation that can justify building the first generation? With these trends (reasonable or not), identify National's decision strategy based on the results of the first generation.

Suggestions to the Student

1. There are a number of assumptions concerning conversion and reconversion costs for the Joliet plant that have to be made.

2. More importantly, the extension of growth patterns into the second generation is critical. The assumption of linear trends may be sufficiently accurate for the first generation, but this assumption becomes less tenable as production grows.

3. It will be easier to keep track of the possibilities, if you draw a decision tree.

Case 23

Washing Away

Seaview is a small resort community on the Gulf of Mexico. Blessed with nice beaches and a good location, Seaview has grown rapidly over the last decade. The growth has not been explosive, so that Seaview has maintained a semblance of "good taste."

Seaview's city engineer, Ramon Martinez, had a number of sleepless nights during Hurricane Harvey last year. His biggest concern was a seawall that protects the first row of hotels, motels, inns, and restaurants from waves and storm tides. The seawall withstood the onslaught, but it was close. Ramon vowed to analyze the city's vulnerability along the seawall before the next hurricane season. He has gathered a first cut at the data, and it is now time to begin the analysis.

The seawall was designed to withstand a 50-year storm with an additional cushion through safety factors. Analysis has shown that Harvey was a 40-year storm for Seaview. Now after 20 years of storm damage, minor maintenance, additional data gathering, and improved design and modeling capabilities, the seawall seems to still be matched to the 50-year level. Barring destruction by a major storm, the seawall with proper maintenance and repair seems likely to last for a century or two.

During the same 20 years, the growth rate has averaged 4% annually (adjusted for inflation), rather than the 1% that was originally expected. The consequences of a large storm have increased with the rapid growth. Ramon believes that the 50-year standard is probably inadequate for the larger consequences. This inadequacy will increase as Seaview's growth is not slowing, and the beachfront property that relies on the seawall continues to be the most prized.

Ramon has identified three types of alternatives for the city. First, future risks could be reduced by restricting development along the seawall. In an extreme version this could include condemnation of existing buildings and purchase by the city. Second, the city could mitigate the financial consequences through insurance. Seaview could require that property owners be insured for hurricane damage. Third, Seaview could increase the level of protection by strengthening the existing seawall.

Ramon plans to use an 8% interest rate in the evaluation. This is relatively high, because Seaview's rapid growth has created many needs and overstretched facilities. He is planning on using a long horizon, at least 100 years. The seawall has a long life, and the higher severity storms have even longer return intervals.

As strengthening the seawall is his "natural" reaction to the problem, Ramon started his analysis with it. The seawall is about 2.3 miles long and is located between the beach and the first row of buildings. It is readily accessible from the beach, and the strengthening project could easily be completed during the off-season.

The seawall's size and cost increase with the severity of the storm that it is designed to withstand. Strengthening the seawall from a 50-year design standard to a 100-year standard would cost $3.15 million. Each doubling of the return interval costs another $3.15 million. Rebuilding to the 50-year standard would cost $4 million, and larger seawalls would cost the same $3.15 million per increase in the design interval.

Table 23-1 summarizes the translation of return intervals into probabilities that Ramon plans to use.

Table 23-1	Probabilities for Storm Severity					
Return interval	50	100	200	400	800	etc.
Inverse cumulative probability	.02	.01	.01/2	.01/4	.01/8	etc.
Probability	.01	.01/2	.01/4	.01/8	.01/16	etc.

The expected damages depend on the difference between the storm's severity and the design standard (interval) used. If the storm's severity is less than the design standard, then there is no damage. If the storm's severity matches the design interval, then damages will equal about 10% of the protected structures' value. If the storm's intensity is one interval higher, then damage will approximate 30% of the structures' value. For two intervals higher, the damage increases to 70% of the structures' value and complete destruction of the seawall. For three intervals higher, the first row of buildings would be completely destroyed (salvage value = cost of cleanup). (See *Condominium* by John D. MacDonald). There would still be no damage to structures on the land side of Beach Boulevard, because the four-lane divided parkway and the first row of buildings will act as a bulwark.

The condemnation alternative is politically difficult at best. First, most owners would not want to sell their property. Second, the appraised value of the buildings along the beach is nearly $200 million, and they sit on land worth $60 million. Another $30 million in beachfront land has not yet been developed. This $290 million in property provides only 15% of the property tax revenue that supports Seaview's services, but Seaview cannot afford to "buy it out." It is probably easier for Seaview to acquire the $30 million in undeveloped land through condemnation proceedings than to significantly restrict the value of ensuing development if it remains in private hands. It may also be easier for Seaview to condemn undeveloped rather than developed property.

The head of the city's legal department responded positively to Ramon's query about the city's ability to require "hurricane" insurance. So Ramon conducted a small survey of building owners along the beachfront to check their insurance coverage. After the first fifteen interviews, he called the city attorney and said, "The situation is far worse than we suspected. All of the buildings are insured, but not one policy allows for failure of the seawall. In fact, six of the policies specifically disallow damage if the seawall fails." The city attorney reassured Ramon with the comment that "At least Seaview can't be sued if the seawall fails." Buying insurance to cover the damages for these extreme floods would cost 50% more than the expected level of damages. Seaview could require it and act as a central coordinator in obtaining coverage.

Ramon has asked you to prepare a report for his review, which emphasizes the economic analysis. He also encouraged you to consider the political factors that may dominate the economics. He asked you specifically to include a table summarizing the trade-offs between the four alternatives. (There are two versions of the condemnation alternative).

Option

It is nine months later, and Hurricane Clara has just destroyed Seaview's seawall and beachfront property. Clara was a 400-year storm, and the actions Ramon had suggested were still wending their way through the political process.

Disaster relief funds have been approved, but they are only covering about 60% of the loss for the private property owners. Seaview appears likely to fare somewhat better, as the governor has promised $13.5 million for a seawall designed to the 200-year standard.

The same alternatives as before exist with some modifications in their expected cost. For example, the insurance costs have gone up by a third. On the other hand, condemnation would now only require that Seaview pay for the value of the land, without the beachfront structures.

Case 24

Sinkemfast

by
Herb Schroeder
University of Alaska Anchorage

I. L. Sinkemfast, Inc., is a drilling company that works on Alaska's North Slope. The oilfields have been constructed on ice-rich permafrost ground which is thaw unstable. Since oil is extracted from the formation at 160°F, the various pipelines needed to develop the field cannot be cost-effectively buried. Instead the pipelines are elevated on pilings. Over the next five construction seasons, about 120 miles of new pipelines will be built. About 56,000 pilings with a total value of about $51M will be let in 5 seasonal piling contracts.

Standard piling installation is to auger a hole 6″ larger in diameter than the piling, plumb the piling in the hole, stabilize it with cribbing, backfill the annulus with a sand-water slurry, and then after the slurry has frozen, remove the cribbing. Conventional pile driving with impact hammers does not work in the ice-rich frozen soil, but an engineer known to Sinkemfast has tested a sonic pile driver. The sonic pile driver oscillates the piling at its resonant frequency to liquefy the ice-rich soils at the soil/pile interface. The piling then penetrates the frozen ground. Preliminary testing on 85 piles indicates a man-hour reduction of 80% per piling. The engineer will lease the rig as-is to Sinkemfast for $3M for a 5 year period. However, the oscillator unit has failed on the average once every 17 pilings. The engineer has told Sinkemfast that investing $1M in R&D has a 70% chance of curing the problem.

Sinkemfast owns a fleet of 6 auger-type drills. Each one has a market value of $62,500 and is expected to last three more years, when they will be worn out and worthless. Using this equipment there is a 35% probability that Sinkemfast will win all of the drilling work each season at an 8% profit margin after deducting bidding costs.

Sinkemfast can purchase 4 new air auger drills for $500,000 each. This drill is faster than conventional augers and will give Sinkemfast a competitive advantage. If the new drills are bought, the old drills will be sold. These air augers will have a market value of $75,000 each after five years. Using the new air auger equipment, there is a 45% probability that Sinkemfast will win all of the available drilling work each season at a 12% profit margin after deducting bidding costs.

If the sonic pile driver is operational, Sinkemfast estimates a 90% probability for winning all of the available drilling work each season at a 20% profit margin after deducting bidding costs. The existing fleet will be sold when the sonic driver is operational. The useful life of the new sonic pile drivers is unknown, but it is believed to be longer than the 5-year license period.

The state regulatory agency is issuing new regulations that tighten the requirements to remove all facilities when production is finished. How does this affect this decision?

The cost of bid preparation is $231,000 annually. Sinkemfast uses $i = 10\%$. What should Sinkemfast do and why?

Suggestions to the Student

While a decision tree that analyzed each year (win or lose bid) could be developed, it is much easier to use expected values for annual receipts. This also better matches the accuracy of the data and assumptions. The more detailed tree would be useful in analyzing changes in bidding strategies (winning several years in row implies that Sinkemfast can increase its bid price or losing several years in a row implies Sinkemfast should make a lower bid). A decision tree may still be useful for evaluating R&D on the sonic driver to fix its failure rate.

Case 25

Raster Blaster

by
Donald Merino and Kate Abel
Stevens Institute of Technology

Mr. Hy Eyeque is an electronics expert who has an extensive workshop in his home for "tinkering." Hy has been working recently to develop a "black box" that will convert video output into TV signals, but with a 100% increase in the number of vertical lines per "raster" scan. In his work on this project he has been attempting to use off-the-shelf components as much as possible, both to speed eventual production and to hold down the costs.

Thus far, Hy has succeeded in producing two prototypes or breadboards that work, but they are far too large and produce too much heat to commercialize. To date, Hy's out-of-pocket expenses total $15,550, plus Hy's own labor and wear and tear on his tools.

Over the past few months Hy has been calling on manufacturing firms in the area (Hauppauge, Reno Electronics, Ryan) to try and interest one or more of them in completing the development of a commercial model of his Raster Blaster box. He has also contacted electronics distributors (Arrow Electronics, EMZ Electronics, Bilco) and retailers (Best Buy, Sears, Circuit City) about distributing/selling such a product.

The most serious interest to date has come from Ryan Integrated Products of Long Island City. Ryan currently produces sound and video boards for a variety of brand name companies (3Com, ATI, Matrox, Creative Labs) and makes some of the chips for these in house. Their development, assembly, testing, and packaging capabilities, plus contacts with companies further along in the distribution chain, make them a prime candidate for Hy's product.

In a recent meeting concerning the Raster Blaster at Ryan, Anne Whyte, Ryan's market research manager, indicated that a study had been conducted by Kirby & Shaw, a well-known market research firm, for which Ryan paid the fee of $100,000. The study results showed that there would be a market potential of *at least* 100,000 Raster Blaster units annually nationwide, if the product were appropriately promoted. The management of Ryan were quite excited at this finding. First, they negotiated an annual license fee of $150,000 with Hy. Then they promptly set up a project team to take the development through the next step: model and documentation development. Tom Cable from the model shop was asked to make thirty models of the Raster Blaster, using production parts as much as possible, but using model-shop production methods, which meant building the units by hand. Fred Mertz of manufacturing engineering was asked to draw up specifications, parts lists/numbers, and production methods for the product, should Ryan decide to go into production.

Costs for these pre-production activities are shown in Table 25-1.

Table 25-1   Pre-production Activities

Item	Materials	Labor	Outside Services	Total
Documents		$975	$125	$1,100
30 Demo units	$1,200	$9,500	$450	$11,150

To support the further development of his idea, Hy also worked in a consulting role on the model production. He spent two man-weeks at Ryan.

Once the models were completed and were placed in test locations under the observation of Kirby and Shaw, focus shifted to the market forecasts and marketing plan.

Bill Kimble, marketing consultant, had developed a plan by which Ryan could produce the Raster Blaster and distribute it through the wholesale channel (to firms such as Arrow Electronics, EMZ Electronics, and Ryco) where the marketing costs would be minimal. These firms would, in turn, place the Raster Blasters with retailers, with a typical distributor mark-up of about 33%. The retailers would offer the product through their stores with a mark-up of some 22%. End user prices were estimated in the range of $83.55 to $106.55.

At these prices, Bill estimated that sales would likely start at 30,000 units the first year, with annual unit growth of 100% per year thereafter. Factory prices are expected to decrease 20% for third-year sales, 25% for fourth-year sales, and 35% for fifth-year sales. (Bill billed

Ryan $85,000 for the study, which included art work and packaging design for the new product.)

Results from the testing of the models proved encouraging, as the reliability was excellent and the higher detail of the picture on ordinary TV sets was receiving kudos galore from all of the testers.

Roger Pedaktor, Ryan's CEO, decided to go full speed ahead, and requested an immediate five-year economic analysis using Bill Kimble's sales estimates, including a $60,000 promotion budget for the first year of sales which will increase proportionally with sales volume, and asked Fred Mertz to provide production cost estimates for the analysis.

Fred estimated that, for the production units:

- Labor would cost 95% less than for the models.
- Material would cost 90% less than for the models

These estimates assume the purchase of $75,000 in production and packaging fixtures and an additional $225,000 in test equipment and software.

Use a MARR of 12%.

Questions

1. Find the fixed capital cost for starting the business.

2. Calculate the unit production cost, unit contribution margin, and breakeven volumes for the worst case from the company's point of view (the low selling price) and for the best case (the high selling price).

3. Calculate the annual demand, selling prices, and profits through the planning horizon at both low and high prices.

4. Determine the range (depending on selling price) for the project's present value.

5. Recommend whether to "go" with the idea or not. Calculate the breakeven value at the low price of the data item that you consider most likely to be unreliable.

Molehill & Mountain Movers

Cathy has one year left before she completes her degree in industrial engineering. She is spending this summer working for her family's firm, MMM (Molehill & Mountain Movers). MMM runs a fleet of heavy construction equipment and sells gravel for roadwork from its pit. They are opening a new section of the pit, and they must choose between conveyor and front-end loader systems for loading the trucks. In the past they have used front-end loaders.

The firm's CPA has asked Cathy to analyze the after-tax cost of the two choices. Her task is complicated by uncertainty over the depreciation portion of the tax code. It is up for revision once again (the prolonged business boom has raised the pressure for increasing business taxes). Thus the system may be depreciated under (a) straight line, (b) sum-of-the-year's digits (SOYD), (c) double declining balance, or (d) modified accelerated cost-recovery system (MACRS).

The current tax system does not have a special rate for capital gains, but it may be reinstituted at a rate of 66.67%. The other complicating factor is the effect of inflation, which the CPA said can be assumed to affect all numbers equally—except for tax calculations based on book values. At least the CPA simplified the task by defining the after-tax rate of return as 6%, and the tax rate as 40%.

The CPA asked for a recommended decision based on Table 26-1.

Table 26-1	Cost Summary for Conveyor and Front-End Loader	

	Conveyor	Front-end Loader
First cost	$250,000	$110,000
Salvage value	$80,000	$30,000
Life	15 years	15 years
Operation cost	$32,000	$45,000

Options

1. Compare the equivalent uniform annual costs (EUAC) for the conveyor and the loader assuming that inflation is 0% under four depreciation methods: (1) straight line, (2) SOYD, (3) double declining balance, and (4) MACRS. Do the different depreciation methods have similar impacts on the conveyor and the loader?

2. Assuming that inflation is 10%, use the same four depreciation methods and redo option 1.

3. Graph the EUACs for each depreciation method as a function of the inflation rate, between 0% and 15%.

4. Compare the relative sensitivity of the inflation rate, the annual operation cost, the tax rate, variations in the first cost or salvage, and the choice of depreciation method.

Suggestions to the Student

1. Assume that MMM is profitable so that costs and depreciation are deductible from taxable income.

2. The Excel function that can calculate MACRS depreciation is VDB. This allows declining balance with a switch to straight-line and no salvage value. For the time period arguments see your text. (Excel's DB function makes different assumptions.)

Case 27

To Use or Not to Use?

Beaufort Sea Production Company (BSPC) operates a medium-sized oil field on Alaska's north coast. The field is still producing at its maximum rate, 325,000 barrels of oil per day (BOPD). However, to sustain this rate the company started a waterflood of the reservoir two years ago. Now a capacity bottleneck in the water disposal process is threatening to curtail production.

In waterflooding, salt water from the Beaufort Sea is treated to remove debris, impurities, and oxygen (to minimize corrosion problems). It is then pressurized to 2800 psig for injection into the reservoir, where it serves two purposes. First, it sweeps the oil toward nearby production wells, which increases oil recovery from the swept area. Second, it maintains the reservoir pressure for all wells by replacing the oil that is removed.

The injected water also becomes part of the fluids that are brought up through the production wells. Over time the wells steadily produce more and more water, which must be re-injected. They are now making 200,000 barrels of water per day (BWPD), and they can dispose of up to 380,000. Over the next three to four years, they expect the produced water rate to increase to around 600,000 BWPD before leveling out.

This produced water does replace an equal volume of seawater as the injection fluid. But due to the incompatibility of the seawater and the produced water (mixing causes immediate precipitation of calcium carbonate scale), different pump modules must be used. More pump modules must be added.

The lead time on new facilities is about 2 years, so alternatives to increase the disposal capacity of the produced-water system must be evaluated now. If the capacity is added too late, then oil production rates must be reduced to match the existing capacity. The economic penalties of deferred production are heavy. In fact, if the production is deferred too long, some of it may still be unrecovered when it becomes uneconomic to produce the field.

Fred, the lead project design engineer, has identified two primary options for the expansion. The first and lowest capital cost option is to complete construction of a module started three years earlier as part of the initial water flood facilities. The company had already spent $12.5 million on this module, when a new water flood plan reduced the area to be flooded. This made the third module unnecessary, construction was immediately stopped, and the module was mothballed. The pump and aero derivative gas turbine had already been purchased and the module partially constructed.

Since then, the mothballed module has been stored at the construction site. The design engineers estimate that it would cost an additional $22.5 million to complete, modify, and install it. Modifications include another produced-water tank and booster pumps to supply the water at the proper pressure for the suction of the high-pressure pump.

Management at BSPC sees this as a chance to salvage a useless module. The pump more than meets the requirements, since it has a design rate of 400,000 BWPD at 3375 psig. At the required pressure (2800 psig), it can pump up to 480,000 BWPD.

If not used for this, the module's only value is for spare parts for the two installed units. The book value of these spares (essentially only the pump and turbine, since the module is unusable) is $4.4 million. An extra $1.9 million is required for this option for replacement of the gas generator spare.

The second alternative is to add more pumps similar to BSPC's two largest produced-water disposal pumps. There are eight total. Each pump in the largest pair uses a 4600 horsepower (hp) industrial-type gas turbine to pump up to 85,000 BWPD. Three of these pumps would be needed. Fred's estimate for total costs is $30.1 million.

Fred realized that he needed some operation and maintenance (O&M) costs. Since he has had very little operating experience, he asked some engineers with the BSPC production facility to help. He was a bit surprised with their response. Operations had experienced many problems with the two 400,000 BWPD waterflood modules, especially during the first year. Correcting several manufacture-related problems had helped, but the pumps still did not run as smoothly as planned. Luckily, short interruptions of the injection of "new" seawater have minimal impact on the production of oil.

The production engineers were concerned with using such a large machine where shutdowns impact production greatly and quickly. Initial calculations of "residence times" indicated that there would be less than an hour to respond to an unexpected loss of the pump. With such a short response time, they would have to immediately shut-in a large number of production wells (80 to 90, if it were running at full capacity). They felt very strongly that they would operate the pump at a reduced capacity, probably less than 340,000 BWPD. Even at this reduced capacity, a number of the smaller pumps would be shut down.

On the positive side, they noted that the excess capacity could be useful should they have to shut down one of their smaller pumps for maintenance. In fact, it could actually serve as an on-line spare.

Since the two 85,000 BWPD pumps had only been installed for six months, little O&M data were available. They did know that when these pumps shut down unexpectedly, the operators only needed to cut back the wells with the highest water production rates. No wells had to be completely shut-in.

The addition of three more small pumps would give them eleven pumps with no full spares. With that many pumps, the chances of one or more being down was substantial. They estimated that the more numerous "minor" cutbacks in production would about equal the shut-in production from a loss of the bigger pump. There is only a negligible difference between the quantities of "deferred" oil for the two options.

To give Fred some numbers for his analysis, the production engineers roughly estimated the O&M costs of each machine. They estimated that the operators spend about half an hour each day conducting routine checks on the large waterflood machines. They figured the smaller disposal pumps take only about 20 minutes per day per machine. They also informed Fred that, when estimating their engineering projects, they generally use $150/hour for operator man-hours, which includes all associated overhead and burden. In addition, they have been given guidelines that indicate that an 8% discount rate should be used for any economic analysis.

Based on the last year, they estimate that routine maintenance on the large pump system will be about $65,000 per year. This includes normal preventive maintenance. Because of their smaller size, they figure that the preventive maintenance on an 85,000 BWPD pump will only cost about $25,000 per year.

Periodic major overhauls are required. In the large aero derivative turbines (used for the 400,000 BWPD modules) the manufacturer recommends replacing the gas generator every three years at a cost of $250,000. The industrial-type turbines have a longer overhaul interval, six years, with an expected cost of $75,000 per overhaul.

134

Major overhauls are needed for the pumps every five years. The large pump overhaul costs $80,000 and a small pump $30,000, exclusive of routine and preventive maintenance. The control systems need revamping every 10 years. Again, the large system is more expensive at $50,000. The smaller systems would each cost about $20,000.

With so much horsepower involved, fuel costs matter. BSPC pays $0.75 per thousand standard cubic foot (10^3 SCF) for their fuel, which has a heating value of 900 BTU/SCF. The aero derivative turbine at 8100 BTU/hp-hour is much more fuel-efficient than the industrial-type engines planned for the small pumps (9250 BTU/hp-hour). The small pumps would require the full 4600-hp rating of their turbines to put out 85,000 BWPD. Based on the pump curves for the existing water flood modules, the large pump would require 18,650 hp to pump 340,000 BWPD.

Another concern of the production engineers was freeze protection when the large pump is shut down. Depending on duration and the time of the year, displacement of oil with methanol might be required. (Below-freezing weather occurs nine months of the year). In addition, the frequent shutdown of equipment tends to increase repair costs for the wellhead chokes.

Freeze protection and extra maintenance costs due to shutdowns of the large pumps would be on the order of $350,000 per year. Since past experience has indicated that wells would not be shut-in when a smaller pump is lost, no freeze protection costs were estimated. Freeze protection costs for the water injectors themselves were not included. The original seawater pumps can maintain sufficient flow to each injector to keep the lines from freezing.

Fred must put all of this information together and make a recommendation to the company management. Should he recommend use of the existing module?

Option

Even though the value of the uncompleted waterflood module has been declining over time, BSPC has not yet been able to take a tax deduction for it. BSPC cannot depreciate the module until it enters service, and the company cannot simply expense it until it has actually begun to dismantle it.

Furthermore, if the module is not utilized for the produced water disposal, all investment tax credits originally taken must be "given back" to the government. This applies to any booked costs not officially transferred to operations. It excludes costs such as the book value of the spare parts that must be depreciated along with other capital equipment when they are placed in service.

The expenditure patterns expected for the two options are as shown in Table 27-1.

Table 27-1	Spending Patterns			

		Prior Years	Year 1	Year 2
	Large pumps	12.5	12.5	11.9[1]
	Small pumps	4.2[2]	15.1	15.0

---

[1] Includes cost of replacement gas generator.

[2] Book value of spare parts transferred to operations warehouse. The pumps are assumed to be placed in service in year 3. The company pays a combined state and federal income tax of 48%.

Olives in Your Backyard

by
Daniel Franchi and M. Lee McFarland
California Polytechnic State University – San Luis Obispo

You may buy an olive farm, where the olives will be sold as olives or olive oil. The trees are planted on 15 acres of the 40-acre parcel. The farm's cost is $500,000, which includes the following assets:

- 3000 olive trees, approximately 4 years old, value = $50 per tree
- 2 wells, value = $12,000 per well
- 1 solar powered pumping system, value = $4000
- 1 large barn and one small out building, value = $65,000
- About 3 miles of fencing, valued at $4 per foot
- About 2 miles of dirt roads, worth $10,000
- 5 gates, worth $1000 each
- 20,000 feet of irrigation hose, at $.20 per foot
- Underground piping for 15 acres, worth $15,000
- Storage tank valued at $3000

For this analysis assume that all depreciable assets (including future purchases) have 7 year recovery periods for MACRS. Assume that existing assets are depreciated from their initial basis over the full-recovery period. Note: the value of the raw land, net after subtracting the value of above assets, is **not** depreciable. Additional investments (assume EOY for analysis purposes) include:

- Year 1: tractor at $25,000 and storage bins at $5000.

- Year 2: fencing at $8000 and $10,000 to connect to local electrical grid with electrical operating expenses of $1200 per year.

- Year 3: added water storage at $9000.

Assume well repairs of $1000 will be necessary every year. Annual operating and maintenance expenses for the orchard are about $20,000, but could vary by up to 25% in either direction. Picking costs are about $500 per ton. Once the olives are picked, they must be pressed into oil. The olives must be transported to a press, at the cost of $100 per ton. Charges for the use of the press are $250 per hour, and the press can handle ½ ton per hour.

Expected production output is about 5 tons of olives per acre. Production will ramp up from 30% of output in year 1, to 60% in year 2, to 100% in year 3 as the olive trees mature. Depending on rainfall, production can vary from 50% to 125% of the expected yield.

The olive crop may be sold either as olives or as oil. For purposes of this analysis, assume that the olives will be pressed into oil. The pressing process yields about 50 gallons of extra virgin oil per ton of olives. Extra virgin olive oil wholesales for about $50/gallon. The price range in the past year has been from $40 to $60. About 20% of it can be retailed at farmer's markets for about $20 per ½ quart. These sales have expenses of $6/bottle plus olive oil.

The income tax rate is 30%. The farm is expected to sell for $800,000 in 10 years (after all depreciation recapture and capital gain taxes.) You are uncertain of the exact interest rate that should be used, but 10% sounds like a good starting point.

Deliverables. Recommendation with engineering economic analysis, any suggested alternatives, assumptions made, and risks considered. At a minimum, it should include:

- 10-year cash flow projection, utilizing sensitivity analysis for prices, costs, and yields
- Calculate an internal rate of return before and after taxes
- Any risks associated with the salvage value assumption
- Graph of present worth versus three important uncertainties, such as price, production, salvage value, etc.

Option

Use more realistic and detailed recovery periods (tractors are 5-year property, trees are 10-year, roads are 15-year, and farm buildings are 20-year). Assume that the appraised and sale value of the land with trees and the property increase to $700,000 and $950,000 respectively before taxes, that the capital gains rate is ½ the normal tax rate, and that depreciation recapture is computed assuming that the value of depreciable assets equals the sales price minus the land's appraised value.

Case 29

New Fangled Manufacturing
by
M. Lee McFarland and Daniel Franchi
California Polytechnic State University – San Luis Obispo

You are the engineering representative on a team for a new product introduction. The proposed manufacturing process uses a semi-automated machine along with people.

Components for each unit of the product cost $8. The semi-automated machine costs $1,500,000, and it has a 7-year MACRS recovery period. The salvage value is $0 for this specially designed machine. This machine can manufacture 175 finished parts per hour.

Table 29-1		Production Volume (1000's)							
Year	1	2	3	4	5	6	7	8	9
Volume	195	275	385	550	625	695	630	550	295

The normal manufacturing operation runs 8 hours per shift per day. Initial production would begin with one shift, 5 days a week. The machine placed in the facility must support this plan. Each working year has 50 weeks (250 regular working days) to allow for vacations. The total labor cost is $50 for each regular time hour that the machine operates and $65 for overtime (these costs include benefits).

Assume that employees can be shifted between production of this new product and other products already in manufacturing. This assumption means that this product is charged with only those hours used and not with one or two full 8-hour shifts.

The production operation can operate a maximum of 8 extra hours/week, if needed to meet the demand without adding an extra shift. This may be a 6th day or some hours added at the end of the regular shift. Many employees like to earn "some" overtime. Thus, the overtime option is more desirable than adding a second shift if overtime can meet the demand.

Other required information:

Corporate MARR	12% (after tax)
Cost of borrowing	9%
Effective tax rate	35%
Maintenance cost	12% of raw material cost
Overhead cost	2% of raw material cost
(utilities, supervision, marketing, etc.)	

The decision of whether to release the new product into production requires answers to the following questions:

- What average selling price of the finished product would be required to yield a 20% after-tax rate of return?
- Is the NPV more sensitive to changes in raw material cost or changes in selling price?
- Is the IRR more sensitive to changes in raw material cost or changes in selling price?
- Do variations in the machine's cost have a significant impact on the IRR or NPV?

Case 30

Supersonic Service?

by
Joseph C. Hartman
University of Florida

The Concorde

The Concorde is the only supersonic jet ever to enter commercial airline service. It was developed, at the cost of £1.134 billion (British Pounds), through a partnership between the British and French governments by firms that are today part of European Aeronautic Defense & Space Co. and Britain's BAE Systems PLC. The program produced 20 aircraft with the first 4 classified as "pre-production" that never entered regular service. The other 16 aircraft were produced at the cost of £654 million (M). The first aircraft from each of the two production lines were used for testing and also not put into service. Of the remaining 14 produced (between 1976 and 1980), 5 were sold to British Airways and 4 to Air France for £23 M each. (An additional £71 M was paid for spare parts and technical support.)

The Concorde's appeal was its maximum speed of Mach 2.04 (about 1350 miles per hour). It had a maximum range of 4500 miles, but only 3700 when fully fueled with passengers and cargo. Fuel usage was 302 gallons/hour when idle, 2885 gallons/hour normal flying, and 6180 gallons/hour when moving from subsonic to supersonic (about ten minutes per flight). Because of its high fuel consumption, over 75 orders from various airlines (including Eastern, United, American, and TWA) were cancelled when oil prices rose as a result of the oil crisis of the 1970s. The remaining aircraft were sold for the meager price of £1 each, with 2 going to British Airways and 3 to Air France.

In 2003 Air France and British Airways retired the Concorde. British Airways had 5 aircraft in service, 1 out of service, and 1 being used for spare parts. The airline had spent £75 M to upgrade the 5 aircraft as a result of safety issues. Of Air France's 7 aircraft, 4 were in service, 1 crashed, and 2 were being used for parts. They spent a similar amount to upgrade their 4 serviceable aircraft. All aircraft that were in service were donated to museums throughout the world.

Supersonic, or Near Supersonic, Again?

In March of 2001, Boeing announced the development of the Sonic Cruiser and diverted $4 billion (B) in funding from the development of an extended 747. The Sonic Cruiser was to travel just under the speed of sound (Mach 0.98) to avoid the noise pollution of sonic booms. This would cut travel times by nearly 20% over conventional aircraft, which travel at Mach 0.8. Designed for 200 to 250 passengers, it would have allowed point-to-point operations rather than the hub and spoke route design required for large aircraft like the Airbus 380. However, Boeing scrapped the program in December of 2002 as the plane's expected high operating cost did not appeal to potential customers.

Questions from the Developer's Perspective

Although development was stopped, it was expected that the Sonic Cruiser would cost $10 B to develop over a number of years. Consider the following:
1. If development costs are evenly spread over five years, each plane costs $280 M to build and sells for $300 M, how many planes must be sold each year for 10 years (following development) in order to achieve an 18% annual return? Assume annual cash flows and each plane is sold in the year it is built.
2. If the market can only bear 100 planes per year over the 10-year span, what is the minimum selling price?

Questions from an Airline's Perspective

The costs to operate and maintain the Sonic Cruiser are expected to be similar to other conventional two-engine aircraft. A 777-300ER carries about 365 passengers, while the Cruiser was designed to hold 225 passengers. Given these numbers, consider the following questions:
3. If the average one-way fare from New York to Tokyo is $825 on the 777, what is the required average fare on the Sonic Cruiser to achieve similar returns? Assume both planes fly at a 70% load factor (30% of the seats are not sold).
4. What is the tradeoff between the required fare and the load factor (between 50% and 100%) for the Sonic Cruiser?

5. The flight is expected to take two less hours in the Sonic Cruiser (roughly 11 hours as opposed to 13 hours). This will allow the Sonic Cruiser to make a round trip each day (with schedule revamps), while the 777 would require two aircraft to meet the same schedule. What is the economic advantage of buying the Sonic Cruiser?

References

Avery, S., ''Boeing to abandon plans for super-jumbo: Major shift in strategy; Mid-sized plane will travel near speed of sound,'' *Financial Post*, March 30, 2001, p. C01

Gunter, Lori, "The need for speed: Boeing's Sonic Cruiser team focuses on the future," *Boeing Frontier*, July 2002

Keller, Greg, "Air France to save up to €50M from Concorde mothballing," *Dow Jones Newswires*, April 10, 2003

Michaels, Daniel, "British Airways will retire Concorde amid restructuring," *The Wall Street Journal* (www.wsj.com), April 10, 2003

Stone, Rod, "Plane makers, airlines in no rush to replace Concorde," *Dow Jones Newswires*, April 10, 2003

Taylor, Alex, III, "Boeing's amazing Sonic Cruiser: It was supposed to change the way the world flies; Instead the world changed," *Fortune*, December 9, 2002

Wiggins, J., ''Boeing abandons plans for supersonic growth,'' *The Financial Times*, www.FT.com, December 22, 2005

Wilson, J., "Boeing's Sonic Cruiser skirts the edge of the sound barrier," *Popular Mechanics*, October 2001

Freeflight Superdiscs has earned a 10% market share for its version of the ever-popular Frisbee. But demand for their product has stabilized in the mature phase of its life cycle, and they are now considering cost-reduction strategies to increase their profits.

The equipment used in one stage of Freeflight's manufacturing process is on its last legs. Three options have been identified for its replacement. The first option, A, is a modernized version of its current equipment and is a relatively labor-intensive approach. The second option, B, replaces the operator with computerized controls. Option B has higher capital and maintenance costs and lower labor costs. The third option, C, adds computerized controls to another process and interconnects the controls, so that waste heat from the other process can replace the natural gas currently used in this process. This option also requires connecting the two processes with insulated pipes and a heat circulating system to salvage the waste heat.

Option A and the current process require a full-time operator (annual cost of $27,000), as well as $3000 annually for operations and maintenance and $13,000 for energy. The machinery itself costs $20,000 and is expected to have no salvage value whenever it is disposed of. With an overhaul every fifth year (at a cost of $8000) the equipment should last 20 years.

Option B reduces the labor cost to $11,000 annually, but the overhead and maintenance increases to $9000 annually. The energy costs are unchanged at $13,000 annually. The cost of the equipment and controls increases to $75,000, but it still has no salvage value whenever it is disposed of. The equipment should still last 20 years, but the cost of the 5-year overhaul increases to $17,000.

Option C reduces the annual energy cost to zero, but the operations and maintenance (O&M) and labor costs increase slightly to $10,000 and $12,000 respectively. The additional controls and piping bring the total system cost to $170,000, and the 5-year overhaul costs to $25,000. The system life and salvage pattern are unchanged.

If the process will be used for at least 5 years and possibly 20 years, and the rate of return is between 6% and 20%, which process should be chosen? Which sources of uncertainty are most likely to change your recommendation as compared with a base case of 10 years and 10%?

The wholesale price index (WPI) is expected to average 6%, and energy prices are expected to inflate at 9%, while labor costs (direct and O&M) due to productivity improvements are expected to inflate at only 4%. Overhaul costs are split between parts and labor and are thus expected to inflate at 5%.

Options

1. Analyze the problem after taxes. Consider after-tax rates of return between 3% and 12%, with a base case of 6%. For estimating taxes, use 5-year straight-line depreciation and assume a combined federal and state tax rate of 40%.

2. Rather than using a straight-line approximation use the relevant 5-year depreciation schedule.

3. Simplify the problem by ignoring inflation.

Suggestions to the Student

1. The financing of the project does not have to be explicitly treated. As a result, only the differential inflation due to energy and labor price trends must be considered. This differential inflation represents a geometric gradient, and it is combined with the minimum attractive rate of return in an equivalent discount rate.

   The rate for labor is?
   The rate for energy is?

2. Overhauls occur infrequently, so they are easier to deal with individually. Replacements and salvage values can also be treated individually. Write the cash-flow equations for the PW of Option A, Option B, and Option C.

3. By equating the PWs of each option, it is possible to calculate breakeven values for A vs. B, B vs. C, and A vs. C. Each value for the life of the equipment will produce a value for the rate of return for each comparison. This comparison can be done three ways:

   - The easiest way is to pick a combination of life and discount rate, and then to calculate the three present worths. This can be tabulated.
   - The second approach is to pick a base case for each value, and then vary the other value to calculate PWs for A, B, and C. This can be tabulated or plotted (for an example, see Figure 1 in Chapter 2).
   - The third approach is to write the three comparison equations. When plotted on a graph of life vs. rate of return, these curves will divide the graph into areas. For each area, an option will be preferred.

Case 32
Mr. Speedy

Mr. Speedy is a heating and air conditioning repair business that was established 23 years ago by George Moustakis. For the first 15 years Mr. Speedy grew steadily, but then George decided that the business was as large as he could successfully supervise.

Today the business revolves around 20 vans that are on the streets, and another four for backup in the shop. The 20 vans are not all out at once as there is day, night, and weekend coverage using 32 technicians. Each technician is assigned to a van, and each van has only one or two technicians assigned to it.

George has tried using a smaller pool of vans, which requires rotating them among the technicians. He found that this radically increased his costs. First, the vehicles were treated more like somebody else's problem. The drivers were harder on the vehicles, and they did not communicate as well with the mechanics. Second, and more importantly, restocking the truck at the end of a shift was sometimes slipshod. Now the technicians are very consistent about restocking the parts used during the day, when they still have paperwork on what was done. Otherwise, they may be short the next day. The technicians receive a completion bonus, which may equal their normal salary. Thus, interrupted jobs that require a return to the warehouse are the bane of the technicians.

The average age of George's fleet of vans has crept up and is now 4 years. His vans generally last 7 years before they are abandoned. He is getting complaints from the technicians and the mechanics. The vans are spending more time in the shop. Most of the problems can be dealt with after the end of a shift, and only a few interrupt the work of the technicians. George has asked his shop manager and his bookkeeper to analyze the economics of van replacement using a 10% interest rate for the time value of money. Their responses are the two memos that follow this introduction.

These vans are somewhat special. After they are purchased from a dealer, another vendor installs a van liner designed for storage of parts. Then the van is completely stocked. New, the vans cost about $14,000, the liners add another $4000, and the truck's mini-inventory costs another $5000. When a truck is retired, its inventory can be transferred, but its liner is worthless.

The annual maintenance costs for the vans start at $500 per year and increase by $200 each year thereafter. As the vans age, there begins to be more of an operating cost due to missed calls. These costs begin at $250 and increase by $750 per year thereafter.

Options

1. Assume a 40% marginal tax rate for combined state and federal income taxes, and use a 6% after-tax interest rate. Ignoring capital gains and investment tax credits, does your recommendation change?

2. Focus on the total number of vans and what changes in the replacement schedule would be necessary to change the number of backup vans. Which parameters are most critical in making this decision?

To: George
From: Igor, Shop Manager

In analyzing our van replacement problem, I followed an example that appeared in last month's *Fleet Manager.* I recommend replacing the vans after 5 years (two years sooner than we do now). I also recommend that we continue to retire any van within a year of retirement (now 4 rather than 6 years) when a major repair becomes necessary. I asked Vincent what interest rate to use, and he told me 10%. My calculations follow.

Year	Market Value	Drop in Value	Interest on Salvage	Maintenance Cost	Total
0	18,000				
1	10,000	8,000	800	500	9,300
2	7,000	3,000	300	700	4,000
3	5,000	2,000	200	900	3,100
4	4,000	1,000	100	1100	2,200
5	3,500	500	50	1300	1,850
6	3,000	500	50	1500	2,050
7	2,500	500	50	1700	2,250

To: George
From: Vincent Cash, Bookkeeper

In analyzing our van replacement problem, I used generally accepted accounting principles. I recommend continuing to replace vans every seven years.

The only reason I do not recommend keeping the vans longer is increased repair (not routine maintenance). For each year longer we keep a van, it spends half a day more per month out of service. With an average age of 4 years, the 20 vans average 40 days per month of repair work. If this were level, we would not have a problem, but it is not. We have to juggle the schedule, when we need an unavailable fifth backup van. This would be worse if we delayed replacement.

Igor should be able to give a better estimate. But, we could assume that every 10 days per month of repair work requires another backup van.

Under the current tax law, the schedule for 5-year property allows us to depreciate using the following percentages (20, 32, 19.2, 11.52, 11.52, 5.76%). The vans cost us $18,000 initially.

Year	Book Value Year's Start	Depreciation	Interest on BV	Operating Cost	Total
1	18,000	3600	1800	250	5650
2	14,400	5760	1440	1,000	8200
3	8,640	3456	864	1,750	6070
4	5,184	2074	518	2,500	5092
5	3,110	2074	311	3,250	5635
6	1,037	1037	104	4,000	5140
7	0	0	0	4,750	4750

Case 33

Piping Plus

Curly, Maureen (Mo to her friends), and Larry founded CML Mechanical Engineers nearly ten years ago. The firm has grown substantially, and it now employs 28 engineers. CML emphasizes industrial and commercial work related to the construction of facilities. For example, they have designed heating and ventilation systems for factories and shopping malls, piping for refineries and chemical plants, and even a treatment facility for oil tanker ballast water. Obviously, they have undertaken many other jobs over the years, but these represent their market niche.

When CML first started business, most of the work was subcontracts for heating and ventilation systems for individual buildings. The jobs are now larger, more complex, and technically far more sophisticated, but CML still generally works as a subcontractor to another design firm. None of the founders, who are still the owners, want to diversify into other branches of engineering or into more general construction or into project management. They enjoy being in close contact with the technical details, even though they do relatively little design work themselves.

Virtually all of the analysis and technical drawings are done with an ever-increasing array of software packages. However, there are beginning to be problems in interfacing with the project management software used by some of their clients. Furthermore, they would like to link the firm's accounting and time tracking and billing software with the firm's software for design, drawing, and project management.

They believe that this integration will allow automatic tracking of time spent on work packages, progress on contracts, and billing breakdowns by project type, client, and employee. This will in turn support more accurate estimating for bidding on future work.

It appears that software packages costing about $25,000 initially are needed, and then there would be a 15% annual fee. This fee covers service, answering questions, and periodic updates. Another $30,000 would be needed for a contract with a software integration firm. Initial training of the firm's employees is estimated at $12,000, and about one-sixth of that as an annual expense. From looking at their history of software usage, they estimate that the life of this "generation" of software will be about 5 years.

Curly was assigned the job of estimating the value of having the integrated software, while Mo and Larry examined possible sources of financing. Should the firm get the software? If so, use their three memos and the firm's financial statements to define an acquisition plan. This plan should identify the timing of the purchases and the source of funds for the purchase. What is the rate of return of your proposal?

Suggestion to the Student

If enough jobs are lost, the firm may have to shrink; but the danger of this can be judged by comparing lost billings to annual billings. The income statement can also be used to estimate the average contribution to profit and overhead from billed projects.

To: Mo & Larry
From: Curly
About: Projected Financial Impact of Integrating Data Flow between Packages

Last year we lost four jobs with estimated billing of $45,000 because we could not meet client data flow expectations. We bid on and received jobs that covered 60% of these potential billings, and we were only out-of-pocket about $6000 in unbillable wages. Even this time was productively spent on an internal short course on changes in OSHA standards.

I'm guessing, but I expect to lose two more jobs each year that we are without the integrating software. It will also become increasingly difficult to obtain other work. I would guess that a 10% to 20% drop in replacement billings would occur each year. Thus, next year I would expect 6 jobs to be lost, and that we would only be able to replace 40% to 50% of the billings.

Since we do not really want to expand, I've not analyzed this as a problem in attracting new business. But we certainly do not want to reduce the scope of our operations, so I believe we should purchase the software.

To: Curly & Larry
From: Mo
About: Alternative Financing of Possible Software Purchases

We all know that we've cut our margins to the bone on last year's bids. I still believe this was the right response to the downturn in construction in our part of the oil patch. We have managed to stay busy, but last year's profits and this year's projections are essentially zero. If it weren't for the payments on our building and the depreciation on it, both our cash flow and profit pictures would improve.

You know I hate to borrow money, so that only leaves me with the option of bringing new stock into the business. We could attempt to bring in a fourth partner, but I would rather sell stock to some of our principal engineers and maybe even other long-term employees.

Such a plan cannot be set up as profit sharing—we have none to share. Thus the most reasonable approach I see is that qualified employees (maybe everybody) be allowed to take part of their pay as stock. Then CML would match on a one-for-four basis. Thus someone who set aside $2000 in income over the year would be credited with $2500 in stock at the end of the year.

I would guess that about half of the 40 employees would participate, although most would only make minimal contributions—at least at the beginning. They might average $100 to $200 each per month at the beginning. Thus, this plan could generate as much as $4000 per month.

As I've thought about this idea I have become enthused because of the non-financial possibilities. If we could tie our best engineers into the long-term success of the firm, there might be less turnover. We do not have a serious problem. But, like the other design firms, we seem to keep our engineers for only two or three years (in most cases). I'd also like to reward people like Marcie who have been with us since the beginning.

Thus, if this is set up as a share-the-reward and share-the-risk plan, I feel it can improve both morale and productivity. At the same time, it should allow us to buy the new software, as well as providing a structure to lower our costs if business gets even worse.

To: Curly & Mo
From: Larry
About: Bank Financing for the Software Programs

The people at First Loutex have lots of faith in our ability to pay back any loans, but money is tight right now—so they want a pretty steep interest rate. They're also concerned about satisfying the auditors that any loans are sound. Consequently, they will probably insist that we use our equity in the building as collateral for any loan with a term over three years.

We could switch to a new bank, but we know and like them and they know and trust us. It's just that with all the bad energy loans floating around and uncertainty in real estate, they have to be very cautious now. Maybe they'd relent some if we pushed, but I just don't know.

Anyhow, they offered us up to $70,000 at 15% over three years or less. If we want more than three years, it'll cost us an extra $4500 in fees for securing the loan with our share of the building. They also want another 1% in interest.

I've attached in Tables 33-1 & 33-2 the income statement and balance sheet from last year.

Table 33-1   CML Income Statement

INCOME STATEMENT		Year Ended December 31
	Last	**Current**
Billings	$4,850,000	$4,820,000
Billable salaries	3,010,000	3,120,000
Gross Profit	1,840,000	1,700,000
Selling, general, and administrative expenses	1,250,000	1,230,000
Operating Profit	590,000	470,000
Depreciation Expense	(580,000)	(525,000)
Profit Before Taxes	10,000	(55,000)
Taxes on income	4,000	0
Net Profit	$ 6,000	$ (55,000)

Table 33-2   CML Balance Sheet

BALANCE SHEET		As of December 31
Assets	**Last Year**	**Current Year**
Current Assets		
Cash	$ 200,000	$ 180,000
Marketable securities	50,000	50,000
Accounts receivable	1,390,000	1,440,000
Inventory	50,000	100,000
Prepaid expenses	20,000	25,000
Total current assets	$ 1,710,000	$ 1,795,000
Property, plant, and equipment	4,350,000	4,800,000
less: Accumulated depreciation	1,760,000	2,380,000
	$ 2,590,000	$ 2,420,000
Total assets	$ 4,300,000	$ 4,215,000
Liabilities and Owner's Equity		
Current liabilities		
Accounts payable	$520,000	$540,000
Taxes payable	80,000	85,000
Bank loan payable	80,000	75,000
Other payables	80,000	130,000
Total current liabilities	740,000	830,000
Mortgage payable (interest at 9%)	1,320,000	1,200,000
Capital stock plus paid-in capital		
(140,000 shares outstanding)	1,400,000	1,400,000
Retained earnings	840,000	785,000
Total liabilities and owners' equity	$4,300,000	$4,215,000

R&D Device at EBP

Two years ago East Beach Products (EBP) designed and built in-house an R&D device to measure a critical parameter in evaluating new designs. Based on lessons learned, the testing engineers have found a generic device which can be modified to measure the critical parameter and which eliminates most of the operating problems inherent in the existing design. You have been tasked with evaluating the device's replacement in a time of tight capital and operating budgets.

The old device was built for $100,000 and currently has a book value of $48,000. The useful life of the device was estimated at 8 years. The device has no market value but can be scrapped and cannibalized for parts/computers/controls valued at $40,000 today and $5000 less in any succeeding year. Currently, the device costs $32,000 per year to operate (labor and material usage) and another $15,000 per year to maintain. Operating costs are increasing at the rate of $3200 per year (mostly in higher material usage rates) while maintenance costs are increasing at $2300 per year.

The new device will cost $125,000 after modification by the supplier. The new device is also projected to have a useful life of 10 years. The device is expected to have an operating cost of $26,000 for the first year which increases by 10% each year. The maintenance costs for the device are estimated at $8000 per year and are expected to increase at the rate of $1500 per year. The salvage value is estimated at $98,000 at the end of the first year and $8,000 less each year thereafter.

What is the economic life of each device? What is your recommendation if the device will be needed for a long but indefinite period into the future?

EBP uses a before-tax MARR of 10%.

Options

1. What is your recommendation if the device is only used on an R&D project, which should terminate in 6 more years?

2. What is your recommendation if the after-tax MARR is 6%, and the after-tax rate is 40%?

Northern Windows

Harvey Newby recently graduated in mechanical engineering, and was hired by an oil company working in the arctic. He moved north, and like many other newcomers, he was impressed with the number of windows many homes had. Harvey's studies had concentrated on pump and turbine design and operation, but he had taken a course in heat-flow analysis. His main memory was of how incensed his "heat flow" professor had been over the repeal of the federal tax credit for energy conservation improvements. However, he also vaguely remembered the professor's statement that windows transmitted nearly ten times more energy than did walls.

These memories were vague, and he did not worry about following up on them, until recently when he bought a home. He discovered that virtually all homes had the same or more windows than did homes in Southern California. When the real estate agent was pressed for an explanation, she replied somewhat untactfully, "You obviously have not spent a winter inside yet. After the first three months, and for the next four, cabin fever is a big problem. Being able to look outside, and feeling open rather than closed in, both help a lot." His coworkers all agreed with the salesperson, so he swallowed his misgivings and purchased a typical home. It has lots of windows.

However, his misgivings came back when he got a $125 electric bill for October. At this point he called the old owner and asked what the electric bills had been last winter. His summary of the answer is in Table 35-1.

Table 35-1		Last Winter's Electric Bills					
Oct	Nov	Dec	Jan	Feb	Mar	Apr	May
$115	$158	$245	$360	$310	$242	$170	$138

Last week he got two shocks. First, he heard on the radio that electric rates might double because of the expiration of long-term natural gas contracts. Second, he read in the newspaper that the Public Utilities Commission was planning on discouraging electric heating by eliminating the old declining rate schedule for a new flat rate (see Table 35-2).

Table 35-2	Electric Rates	
Old Schedule		Proposed
KWh	$/KWh	New Schedule
0–500	0.115	
501–1000	0.09	$.08/KWh
1001–3000	0.07	
3001–6000	0.055	
6001 & over	0.05	

Harvey cannot afford to sell the home and buy another one, so his only options are to turn down his thermostat, install energy conservation improvements, or switch to a new fuel supply. Luckily, his area of town is scheduled to receive natural gas within the next four years, so he can look forward to replacing his electric furnace. And it is worth looking forward to, since at current prices natural gas costs about 20% as much as electricity for equivalent heating values. But it certainly won't pay to switch to fuel oil for the short-run, and both coal and wood are far too inconvenient.

Harvey paid for a state-sponsored energy evaluation of his home, but there were relatively few suggestions. Harvey has bought a well-built home, so caulking and other minor improvements are of little value. Two improvements seem more substantial. He can add an "arctic entry" to his main entrance and/or he can provide insulation for his windows. Because

162

these two improvements were recommended in the energy evaluation, Harvey can receive a state-subsidized loan to complete them. These 5% loans are available for a 20-year term.

For esthetic reasons, Harvey does not wish to add an arctic entry. This would involve extending his roof and entry, and adding another door to provide an air lock. For esthetic reasons he does not want to replace his current miniblinds with insulated drapes. Thus the option he wishes to consider is adding additional panes of glass to his existing windows.

The payoff of adding panes comes through lowering heat losses due to conduction and radiation. Both of these phenomena can be modeled on a square foot basis, so he believes he can simply choose a single window size to analyze in detail. Later he expects to modify this analysis for window direction (south vs. north), type (opening vs. fixed), and size.

His home has 19 clerestory windows, each of which is 16" ×32" in size. With so many of this size, he has decided to analyze these first. These windows do not open, and they will also be easy to modify. They face south and it is a cold climate, so the net impact of radiation is positive. As a result, Harvey does not plan to consider the various coatings that are designed to reduce radiant heat flow. Since clear window glass has an insignificant impact on radiant heat flow, he is left with a simple problem in conductive heat flow.

He calls the glass company and asks for quotes on 16" × 32" windows. Their response is $8.54 for a single pane and $16.91 for a thermopane unit. He also estimates $5 of wood, paint, glue, etc., would be required for each window. More importantly, he estimates 2 hours of labor for each window. Since this kind of project is a nice break from working and playing at a computer, he values his time at $5 per hour. His windows are already double-pane. Thus if he adds the glass, he will end up with either triple- or quadruple-pane windows.

He has completed the heat-flow analysis and simplified the equations (as shown in Table 35-3) to ease the economic analysis. To keep his home at 70° requires 10,911 degree days per year of heating. *The average yearly outside temperature of 40 °F implies that he must supply 30F° for 365 days.* As long as he wants to keep his home at 65°F or warmer, he must heat year-round (in this northern climate), so adjusting for lower temperatures simply requires multiplying the temperature change by 365 days per year.

Table 35-3	**Formulas for Calculation of Heat Requirements**		

Annual Heat = (degree days) × area × conversion factor × (24 hr/day)

His conversion factor is stated in terms of BTUs per hour-ft^2-F°.

Note that 1 kWh = 3413 BTUs

Number of Panes	2	3	4
Conversion factor	.48	.33	.24

So energy cost savings from increasing the number of panes are based on the difference between the conversion factors.

Since he set the problem up for the economic analysis, Harvey has realized that he wants some help. And you are the lucky person he's asked (you teach engineering economy at the local college). There is quite a bit of uncertainty, and he believes either complicated equations or a very large number of spreadsheets will be required to analyze the problem. Thus he has asked you to determine which variables are the most critical to his decision making and to summarize the relative sensitivity of your recommendations to changes in the variables. He also has asked that this option be compared with the possibility of lowering his thermostat.

In summarizing his economic position, Harvey states that he expects about a 10% return on his investments. As a result he believes the state loan is a good deal, but he is also wondering if the improvements could be justified without it. Harvey also believes the improvement will last as long as the home—probably 50 years, but he does not expect to own it that long. His best guess is that he will sell the home in 10 years.

Options

1. Rather than assuming no inflation, analyze the impacts of inflation in the money supply on the value of the load repayments and energy bills, and varying inflation rates for electricity and for natural gas.

2. Consider the impact of income taxes on the decision. Harvey's marginal tax rate for federal and state income taxes combined is 30%. Assume that commercial financing

would be tied to his mortgage, so that the interest payments would still be tax deductible. (The subsidized loan is not tax deductible.)

Suggestions to the Student

1. The power bills on Harvey's house and the old power schedule can be combined to calculate his average cost of power and his power usage. Then the new schedule provides a base case for his future power costs.

2. A second case is to double these costs, due to the expiration of natural gas contracts. This change is less certain, and its timing even less so.

3. If a horizon shorter than 50 years is used, then there is likely to be some salvage value for the window improvements—the home is likely to sell for a higher price.

4. When natural gas becomes available, the amount of energy lost through the windows will not change. Only its price will. By what factor?

5. The uncertainty of when the natural gas will become available can be limited by two extreme cases—immediately and never.

Case 36

Brown's Nursery (Part A)

Jo Brown's nursery operation has grown from a small herb plot into a thriving nursery business. There are 10 full-time employees and 20 seasonal (part-time) employees. For the last three years taxable income for Brown's Nursery has been steady at $350,000 per year.

While Jo is pleased with the current success of the nursery, she is considering a new contract to supply a local grocery chain with fresh herbs year round. The grocery chain's current supplier is retiring in a year and is planning to sell his business to one of the other grocery chains he also supplies. Rather than assigning a large portion of her current capacity to this contract, she is considering expanding production. Jo has been offered a 5-year contract, starting in 1 year.

To service this new contract without reducing current operations requires purchasing an adjacent piece of land and constructing additional greenhouse space. The property can be purchased for $200,000. The most economical solution to the greenhouse addition is to construct two modular greenhouses for $70,000 each. The modular greenhouses have a useful life of 12 years and a $5000 salvage value. Startup expenses are expected to be $22,500. The purchase of the land, construction of the greenhouses, and startup are projected to require one year. Incremental working capital for this project is $90,000 beginning with startup.

Sales from the contract are forecast at $380,000 each year. Variable costs are estimated at $250,000 the first year, and Jo believes they will decrease at the rate of $5000 per year, as they become expert in growing the new items in the new greenhouses.

Incremental variable overhead for the new space is expected to be $30,000 per year but when the total overhead is re-allocated (based on square feet under glass) the new production will be charged $45,000 per year.

Upon completion of the 5-year contract, Jo believes she can either obtain another contract from this customer, obtain a similar contract from another grocery chain, use the project's assets to meet increased demand for herbs from current customers of the original operation, or dispose of the project's assets. She believes the land can be sold for what was paid for it and each greenhouse is expected to have a market value of $40,000 at the end of the 5-year contract. This decision will be made early enough in the fifth year of the contract to dispose of the project's assets in that year.

The state tax rate is 11%. Jo uses an after-tax MARR of 12%.

Case 37

Brown's Nursery (Part B)

Jo Brown realizes that her initial analysis of the nursery addition in Brown's Nursery (Part A) ignored inflation. She has asked you to reevaluate the project with the following modifications:

1. General inflation for the next five years is expected to average 3% per year.

2. Her contract is fixed—i.e., the income from sales is in actual (or $Year_t$) dollars.

3. The land is expected to appreciate 5% per year.

4. The salvage value is stated in constant-value ($Year_0$) dollars.

5. The working capital needs will pace inflation.

6. Variable costs will pace inflation (as will the annual savings).

7. Overhead costs will inflate 1% less than inflation.

8. Jo's MARR is a market rate.

Is the project viable when inflation is included in the analysis?

West Muskegon Machining

West Muskegon Machining and Manufacturing (WM³) is owned by Wally O'Leo, a degreed manufacturing engineer. Upon graduation from college 20 years ago, Wally went to work for his father in the family business. When his father retired last year, Wally became the firm's manager.

The current products made by the business have a steady demand but little, if any, growth potential. The products are machined using standard CNC machine tools and are assembled using specialty tools and fixtures, most of which were designed by Wally. The business has a good local reputation for quality and on-time delivery. The prices for the product are reasonable, and it is difficult for competitors to compete on price due to WM³'s specialty fixtures. As a result, Wally sees his core business as secure. Wally also sees the business as stagnant and worries that the business has not added a new customer or product in 2 years. With actual taxable income steady at $525,000 per year Wally is concerned that, adjusted for inflation, the business is losing ground. WM³ pays state income taxes at a rate of 11%.

Based on WM³'s reputation, it is often asked to bid on parts and sub-assemblies for the large furniture and automotive parts plants in the region. Wally sees this as an untapped opportunity to add to the product line and grow the business in real terms. Wally has recently received a request for a proposal from one of the area's furniture companies. He has discussed this with the purchasing department of the furniture company, and now he is pondering expanding the business.

The furniture company is offering a four-year contract for 100,000 units per year at a fixed price of $55.50 per unit. WM^3 has sufficient manufacturing and warehouse space to support this project. Wally's people have estimated that the unit cost to produce this item will be $28.50 for materials, $12.50 for labor and $8.50 in variable overhead (60% wages and 40% other items) at today's costs. No additional fixed costs are anticipated although the accounting manager wants to charge the new product a share of the fixed costs based on the square footage this product will need. This amounts to $200,000 per year.

Production will require the purchase of two new stamping presses at a cost of $125,000 each. These presses have a long life and decline in value at about 10%/year. Three dies will also be required at a cost of $36,000 each. Each die, with proper maintenance, is expected to have a useful life of 400,000 units, but they have no value except for this product. To support production, Wally's production planning people are planning on keeping an average of a one-month's supply of raw material inventory on hand. Wally expects to be able to start production next year.

Wally has been assured by the furniture company that they will pay all invoices within 72 hours as long as he meets delivery schedules. The furniture company will expect weekly deliveries equal to 1/50th of the annual demand. To insure timely deliveries Wally plans to keep a week's supply of finished-goods inventory on hand.

Wally has $100,000 on hand to invest in this project. His banker has told him that if invests the $100,000 on hand in this project, the bank will finance the presses at 8% per year with yearly payments over four years with 10% down. The dies can be financed at 8% over three years with 30% down. Inventory can be financed against the company's revolving credit limit at a rate of 10%—paying just the interest each year and using the inventory as collateral until the project's end when the bank is repaid the loan's principal by liquidating the inventory. A "last in-first out" method is used to value inventory.

Wally's real after-tax MARR is 12%. Wally is expecting 3% per year growth in raw materials prices, 2% growth per year in wages. He expects the general rate of inflation to be 2.5%.

Should Wally proceed with this project?

Option

Does financing this project make it more or less attractive (assuming there were funds in the company available to use for this project)?

Uncertain Demand at WM³

The production manager and manufacturing engineer in the assembly department of Western Muskegon Machining and Manufacturing (WM³) have designed a new assembly fixture. The fixture will reduce the assembly time from 33 minutes to 27 minutes, saving $1.92 per unit in direct labor. The assembly department is seeing steadily increasing demands and will soon be looking at overtime or adding more part-time workers.

The fixture will cost $18,000 to fabricate and can be ready in one month. This will place the fixture in service in the middle of the current fiscal year. The fixture will have a ten-year life if properly maintained. Its salvage value will be negligible, as its general usefulness is limited. The fixture is not in the current budget, but both the production manager and the manufacturing engineer believe that the project should be considered for funding out of the budget for contingencies. With wages currently projected to increase at a rate of 4.5% per year over the next six years due to a tight labor market, the project supports the company's goal of reducing the labor content of products as much as is feasible.

In figuring the cost savings, the production manager points out that $88 of overhead is charged for each hour of direct labor. The manufacturing engineer points out that in this overhead charge are the fringe benefits and other payroll-related costs of 38.5%, which are incurred for each hour of direct labor, and that this rate is expected to increase by 1.5% per year.

The new operation will require an additional 100 square feet of floor space, and currently the department pays $18 annually per square foot in overhead charges. The space is available in an adjacent department, which recently reduced its requirements when it did a similar project. Overhead costs are expected to increase at the rate of general inflation (2.5%).

The product has a demand of 2000 units this year (spread equally over the year). There is some uncertainty on the demand over the next five years. Marketing believes that there is a 45% chance that demand will fall off at the rate of 200 units per year until demand ceases after five more years. There is a 25% chance that demand will remain constant until it ceases after 5 more years. There is a 30% chance that demand will be 1000 units for the next two years and 750 for the next three, when demand will then cease.

WM[3] has a real after-tax MARR of 12% with a marginal combined tax rate of 41.3%. Should this project be considered, if budget funds can be made available?

Olympic Bid Perspectives

A local organizing committee (of private citizens) is working hard on winning the right to host the Olympic Winter Games. Much discussion and some debate have taken place on the funding strategies and on local economic impacts. Although the mayor has received many questions about the bid, he has deferred direct answers in hopes of obtaining more information. To date, he has collected three documents that address different perspectives of the Olympic bid.

In hopes of developing a public position, the mayor has asked you (his economic advisor) to review the documents and provide an assessment of the impact to the local economy. He has specifically requested that you provide a discussion of long-term impacts created by the new facilities.

These documents are summarized individually. The first document is the official bid book to the International Olympic Committee. It emphasizes the plans for putting on a successful Olympics—technical, economic, etc. The second document was commissioned by the local organizing committee to demonstrate the positive impact of holding the Olympics. The third document is a "critical" series of questions posed in a lead editorial of the local newspaper.

Organizing Committee Bid

The Local Organizing Committee (LOC) has the responsibility to assure that quality venue sites are provided and to administer a safe and efficient Olympic Games. The proposed budget for this is shown in Table 40-1.

Table 40-1   Organizing Committee Budget ($ millions)

Capital Construction		
Olympic Village	45	
Main stadium	35	
Ski jump	15	
Bobsled/luge	16	
Municipal projects	15	
Other capital	9	
Total capital construction		135
Operating Cost		
Communications	55	
Administration	46	
Security/housing/transportation	35	
Other expenses	23	
Total operation cost		159
Contingency		10
Total cost		**304**
Revenues		
Television	180	
Sponsorships	40	
Local government	40	
Ticket sales	28	
Other revenues	16	
Total revenues		**304**

Existing facilities will be supplemented by substantial new construction. New student housing will be built at the local university for accommodating athletes and team officials.

This Olympic Village is funded $40 million by the state government and $5 million by the LOC. A 20,000-seat capacity indoor stadium, expected to cost $35 million, would be the largest in the state. The budget also includes $15 million for municipal road, parking, and airport upgrades. The remaining $40 million would be spent for miscellaneous construction at other venue sites.

The operating costs are fairly evenly divided between communications, administration, and general support. Communications include publications, staff communication, advertising, and general equipment rentals. Administration includes personnel payroll, data processing, general supplies, and other support items. The remaining $58 million is budgeted for security, housing, food, transportation, venue preparation, and ticketing.

If the contingency funds are not needed, they could be used as seed money for a long-term facility maintenance fund. Current plans are to donate leftover funds to the international committee.

The total costs are offset by expected revenues. Exclusive television rights and corporate sponsorships are the largest and most uncertain revenue sources. As noted previously, the state is expected to fund $40 million to build the Olympic Village. Overall, the revenue estimates have been approached conservatively. Therefore, it is planned that the Olympics will be funded primarily with external monies.

Economic Impact Study

The impact of the Olympics to the local economy will begin a few years before the games and extend for decades afterward. The total impact can be subdivided into the following:

1. Pre-Olympic activities
2. Facility construction
3. Operation of the games
4. General expenditures
5. Tourism impacts

Pre-Olympic activities include trial athletic events at each venue. Eight events (each for four days) at 450 visitors result in 14,400 visitor-days. Likewise, the media and security personnel will begin their stay up to 10 months before the Games, for a total of 456,000 visitor-days (valued at $45 million locally).

The LOC budget includes $135 million for facility construction and $159 million for operating costs. The construction projects will provide local jobs, but some of the materials will be purchased from other areas; thus, the local factor is lower than for the operating costs.

General expenditures by attendees and participants will boost the local service sector. Even though many tickets will be sold to residents, 50,000 non-local spectators are expected to generate nearly 250,000 visitor-days. In addition, over 190,000 visitor-days will result from the 2-week stay of athletes, team officials, sponsors, and VIPs. Overall, $56 million will likely be spent by these people for housing, food transportation, souvenirs, and entertainment.

After studying the current tourism growth rate and annual expenditures, it is clear that any positive impact in this category could provide very significant local benefits. The current tourism business injects almost $300 million into the area, with growth rates over the last 7 years of 5% per year. The media exposure from hosting the Games is expected to boost tourism the first year by an extra 5% and the following year by an extra 2.5% (returning to the current rate of 5% per year thereafter). The net present value of this impact is approximately $500 million.

The total impact is based on the fraction of these gross expenditures that will initially be directed locally. In Table 40-2 this local factor is applied to each segment, and then the added impact of each dollar circulating around the local economy is calculated using a multiplier of 2.6.[1] The local economic impact is projected to exceed $2 billion.

Table 40-2   Economic Impact ($ millions)

Segment	Gross	Local Factor	Local Value	Total Impact
Pre-Olympic activities	45	.9	40.5	105.3
Facility construction	135	.8	108.0	280.8
Olympic Games operations	159	.9	143.1	372.1
General expenditures	56	.9	50.4	131.0
Tourism impacts	500	.9	450.0	1170.0
	895		792.0	2059.2

---

[1] For the Los Angeles Olympics a multiplier of 3.5 was used, but LA is a much larger, much more self-sufficient area.

Editorial Comments

As the business editor for the local newspaper, I believe it is my responsibility to seek more information on the Winter Olympic bid. I would like to ask our mayor and the Organizing Committee the following four questions:

1. What long-term maintenance costs do the new facilities impose on the community? After all, our current convention center and sports arena operate at a loss.

2. Who pays for any revenue shortfalls or facility cost overruns?

3. How do you justify the estimates taking credit for an economic impact created by money that would have been spent here anyway (e.g., student housing costs or spending by local residents for tickets sales)?

4. What would be the tourism impact of negative publicity caused by poor weather conditions or terrorist acts?

Suggestions to the Student

1. There have been some inconsistencies in the treatment of the time value of money. For example, only the tourism impacts specifically are stated in terms of present worth (and year zero is unclear for that). Which costs and impacts should be adjusted up or down? Is this error significant? How should it be estimated?

2. A public facility might show a loss because only part of its benefits can be determined through ticket sales and other direct charges. How does this relate to the editor's first question, or why else is it important?

3. There are no figures included for long-term maintenance and operation of the venues. How could these be guesstimated, do they matter, and are they positive or negative? Are they balanced by other omitted figures?

4. How would you adjust the final impact table to summarize your results to the mayor? The template below (Table 40-3) might be helpful.

Table 40-3   Economic Impact ($ millions)

Segment	Gross	Local Factor	Local Value	Total Impact
Pre-Olympic activities				
Facility construction				
Olympic Games operations				
General expenditures				
Tourism impacts				
Other				

Case 41

Metropolitan Highway

Metronight is the fastest growing city in the country's "Moon Belt," with a growth rate of nearly 11% per year for the last decade. This growth has offered unprecedented economic opportunities both to long-time residents and to newcomers. It has also brought pollution problems, severe traffic congestion, and increased community tensions.

These tensions have come from a more rapid pace of life and from an uneven distribution of the benefits and costs of growth over the communities that make up Metronight. In particular, suburbs, strip development, industrial parks, and shopping malls have grown rapidly. On the other hand, the older residential and industrial areas have not grown, but instead have decayed marginally. During the last year there has been an alarming increase in the unemployment rate within the central city. Even worse, one of the largest industrial plants (9000 employees) has announced that it will close, rather than renovate, its facility within the year unless the city significantly improves the transportation network.

Because of the rapid growth, construction of expressways has consistently lagged behind the level of traffic demand. Commuters to the older industrial core from the newer residential areas face delays averaging about 45 minutes for both the morning and evening trips—in addition to the normal travel time between work and home. More importantly, the city's road system within its older core suffers from aging pavement and bridges. Already, larger tractor-trailer rigs (those over 30 tons) must detour around a weakened bridge that crosses the river that forms the eastern boundary of the older industrial core. This detour adds about 20 minutes to the average one-way trip (they are scheduled to avoid rush hour).

179

A group of leading citizens has put together a redevelopment proposal that combines the efforts of Metronight's government with private sector initiatives. Their proposal would affect property taxes and the transportation network, as well as requiring long-term training and employment commitments from employers. Since it was put together by private parties, Mayor Andrea Fineglass is certain that they perceive its benefits as exceeding its costs.

She is worried that this comes through governmental subsidies. Thus she has asked you (you work in the municipal engineering department) to analyze the proposal from a municipal perspective. Then she will either commit to support the proposal or ask that it be modified to better suit the city.

One direct cost to the city is a proposed moratorium on property tax increases in the industrial core. This would freeze tax rates and assessments for the first five years for industries in the older core area. The booster group has pointed out that private sector improvements are expected to increase the value of the core by 50% with the proposal. They also claim that without the revitalization effort assessments will fall by $10 million per year until they reach half of the current value.

The other large direct cost is the necessity of raising taxes (either a sales tax or the mill rate for property taxes) to cover greater municipal expenditures on the transportation improvements. Calculation of these direct costs is complicated by a large indirect cost—that state and federal assistance for transportation improvements must be dedicated to this redevelopment. Transportation improvements are currently funded 50% federal, 35% state, and 15% Metronight for both normal projects and for this package.

Specifically the redevelopment calls for investment of $75 million over the first three years (currently Metronight averages about $18 million in transportation projects per year with a B/C ratio of 1.5). The mayor has stated that she believes the increased level of funding can be achieved. She expects that $3 million of urgent (B/C ratio of 3.0) projects will have to be funded directly by Metronight for each year of the construction period. Political trade-offs are also expected to reduce general transportation funding to $10 million for the first year after the redevelopment construction. This will increase back to the current $18 million over the following five years.

During construction of the transportation improvements, additional short-term disruptions will occur. The disruptions are likely to increase delays for commuters and business traffic by 25% the first year, by 40% the second year, and by 30% the third year. The final impact should be to halve the commuter delays and to eliminate any detours for heavy trucks.

The transportation improvements will require the acquisition of about $40 million in residential and small-business property. The process of negotiation (and condemnation where

180

required) is likely to add $3 million in legal and administrative costs. About 75% of the property payments seem likely to lead to new construction or renovation in nearby areas. This property will be subject to the same 3% tax rate as the rest of Metronight. (This rate is high because of the decaying core and rapid additions of new facilities). The remaining 25% of the property payments will be used to "retire to Florida" or other options that remove money from the Metronight area.

The industrial core currently generates on a daily basis 40,000 one-way person-trips for its workforce and another 1500 truck trips (although only 10% are above the 30-ton limit for the detour). The annual average wage for these workers is $17,500, although the truck drivers average $25,000.

It appears that this revitalization program would not have augmented the city's growth rate if it had been initiated sooner. Rather, there would have been a different geographical pattern of growth for Metronight. If a large number of the industrial facilities in this older core are relocated, many of them will relocate to other communities rather than to the growing areas of Metronight.

This older industrial core has a total of about 19,000 employees and about $220 million in assessed valuation. Also the mayor, in response to your plea for guidance, has selected a 7% discount rate, a 30-year life, and a value for leisure time of a third of the wage rate. She left you with the problem of estimating any indirect or consequential costs.

Options

1. Include the following variations for a property tax moratorium: (a) as proposed on improvements, but inflation adjustments on current base are allowed; (b) sooner or more rapid adjustments; or (c) on reassessments, but not on rates.

2. Include specific impacts on the city's employment. In the short term, 20% of the additional construction expenditures are expected to swell the income of residents. In the long run, the different pattern of jobs is expected to slow the growth of the city's population. Rather than new residents moving to the suburbs from other localities, more of the jobs will be filled by residents of the city's core (where there is an unemployment rate that is five times the rate for the rest of Metronight).

Case 42

Protecting the Public

The Greenacres City Council has just been informed that the city park may have had a low-grade contaminant included with the latest annual mosquito spraying. The city's health department has been assigned the task of testing for the contaminant and quantifying the health impact of exposure if the contaminant is present.

Even though the department cannot yet quantify the effects, they have identified the causal chain for the health problems. The key link here is burning of contaminated vegetation, which puts the contaminant back in the air as a vapor. This vapor is irritating to the eyes and may cause acute damage to people wearing contact lenses. This is substantial evidence that there is no problem if it is touched or even eaten. Thus the chief danger appears to be campfires, disposal of windblown leaves (which would mix in with local residential area), or the chance of wildfire in the park.

The health department has also verified that sunshine reduces the toxicity for the potential contaminant over a 2 to 3 year period. By year 4 the danger should be gone.

Because of the adverse publicity that the council expects, it has decided that a benefit/cost analysis must be conducted immediately—before the health department can supply any results. In fact, their meeting is tomorrow night, and you (the city's manager) have been ordered to have preliminary results by then.

While they would like to have definitive conclusions, you believe that they will settle for an analysis that quantifies the two worst-case scenarios. One scenario involves removing all vegetation and turning the park into a landfill, while the second scenario involves the maximum health cost in the complete absence of recovery efforts.

The city engineer has dragged together the estimates for the landfill scenario. Here the first step is to remove the vegetation and burn it in a specially controlled environment where the exhaust is passed through a catalyst. Fortunately, the city is in an industrial area, so that facilities for disposal of hazardous wastes are available. If these facilities were not close by, then costs would be even higher than the expected $2 million to $5 million.

The second step is to create a substantial cover by using the site as a temporary landfill. Fortunately, water table and location factors are positive, except that 20 homeowners will have to be bought out through eminent domain. The appraised value of their properties averages $135,400. The city engineer suggests that all of the other costs associated with using this as a landfill can be ignored, since the city would be incurring them at another site in any case.

In about three years the cover would be deep enough and the landfill could be converted back into a park. The park currently emphasizes vegetation, jogging paths, and similar natural uses, but plantings and regrowth are likely to be far too slow. Thus the engineer has suggested using the area for a complex of playing fields for softball, baseball, soccer, and football. This would cut the regrowth time to the minimum and would only require $300,000 for construction.

The planner for the city park department has looked at usage levels for other playing fields within the city, and has used benefit numbers developed for justification of another small group of playing fields last year. From this she has suggested that the playing field complex would have a "net benefit" stream of $125,000 annually after maintenance costs are considered. She also identified 5% as the discount rate that the city council accepted for that study.

The scenario for health impacts if there is no recovery effort is far less defined. Here the difficulty lies in estimating the probability of different kinds of fires, and then estimating how many people might be affected. A wildfire is clearly the least likely; but it could affect the 35,000 people who live in moderate proximity to the park. The fire chief has estimated the normal wildfire possibility at .002, but this may increase to .01 due to publicity and arson.

On the other hand, over a 3-year period, someone is almost certain to build a fire with a stick that has been sprayed. Over 20% of the park's visitors are picnickers, and a quarter use wood, not charcoal fires. This year's leaves are less of a problem, although taking all leaves to the special incinerator is likely to cost nearly $1 million (from close-by neighborhoods, as well as the park).

It is clear that, if the contaminant is there and if you burn it under your hamburger, there is still a sizable chance that you will have absolutely no problem. But, the exact danger is not known. It appears likely that the cumulative danger is related to park usage, which is currently 250,000 visitor-hours per year—growing at about 15,000 hours per year, slightly faster than the city's population.

You know that other uncertainties include the study period and the value of avoiding eye damage. Clearly, an exact answer is impossible. Nevertheless, an answer is required.

Suggestions to the Student

1. For the landfill/sports complex option, is it appropriate to include the benefits of the later years?

2. What study period, fraction of population with contacts, etc., must be guesstimated?

3. The role of population growth is likely to matter only for park usage, and not for potential eye damage. Why?

4. What other options should be suggested, based on the economic analysis? Which pieces of the two options have particularly good or poor benefit/cost impacts?

Bridging the Gap

Stuckagain Heights is a bedroom subdivision of a rapidly growing Alaskan city. It is surrounded by public lands (state park, military reservation, and city park), and its only access is an old army tank trail that has been marginally upgraded. This 3.5-mile gravel road through the city park has 22% grades. The snow and ice season is October through May, so the city's standard for subdivision roads specifies a maximum gradient of less than half that, or 10%. During "breakup" (the spring melt) the gravel deteriorates into mud, and even the large road graders have gotten stuck.

Because of the poor road conditions, this is one of the very few areas of the city that has no mail service, no school buses, and no home delivery of newspapers. This road meets the standard for a "wilderness" park. And in fact, many city residents come to the park for jogging, hiking, skiing, snowmobiling, dog sledding, etc. Thus, users of the city park often jog or park along the road, while residents travel the narrow road at 30 to 40 mph—slowing only for blind curves, for moose, and for accidents.

Unfortunately, major improvements or any realignments of the existing road require the approval of the U.S. Secretary of the Interior. Local environmental activists ensured that subdivision access was specified as a nonconforming use when the land was conveyed from the federal government to the state and thence to the city. They seem likely to oppose any improvements, which will certainly complicate the process of gaining approval from the Secretary.

Nevertheless, mounting traffic levels ensure that the city must face the problem of upgrading the road or building a totally new access. Besides park usage, which has doubled in the last five years as the city's population increased by 60%, there has been a substantial increase in the number of homes in Stuckagain. Ten years ago there were 21 homes, and today there are 73. Traffic surveys indicate that each house generates about 4.8 trips per day. The rate of this growth varies with the city's economy, but zoning and topography indicate that the eventual capacity of the region is 350 to 375 homes.

The residents of Stuckagain prefer development of a new access road, even though it seems likely to be the more expensive alternative. To begin with, they are intimidated by the necessity of obtaining approval from 4500 miles away for the realignments that are essential to upgrading the current road. In addition, many Stuckagain residents are convinced of the basic incompatibility of park use and subdivision access. Since the new road would meet current city standards and since it is somewhat shorter, this road would also greatly reduce the "hassle" and economic costs of travel.

This new access road will pass close to another subdivision, Chugach Foothills, which is completely developed with 600 homes. Their community council violently opposes the new access road. Thus, both "solutions" have some political problems.

The mayor and city manager are caught in the middle. In hopes of generating a factual base, they have asked you to develop specific dollar figures for the new road based on historical analogies. This total cost can be a starting point for evaluating the possibility of upgrading the current road.

One difficulty with the use of analogies is that road construction costs depend so heavily on soil conditions and distances. The availability of on-site gravel can make a fivefold difference in the cost of roadbed construction. The new road passes very near to a site that had been targeted for a gravel pit by the city for 40 years. The mayor claims to be almost certain (90%–95%) that the military will permit use of this site.

The other uncertainty involves route selection for the first mile of the road. If environmental groups accept a route just inside of the city park's boundary, then it will probably be approved by the Secretary of the Interior. This route avoids the cost of crossing environmentally fragile wetlands, which lie just outside of the park. (The route inside the park is also farther from the objecting subdivision, Chugach Foothills). If this route is not accepted, then an additional $0.5 million to $1 million in project costs will be required to mitigate the environmental damage.

Similar projects from the last construction season had per-mile averages of $25,000 for surveying and design, $33,000 for clearing and site preparation, and $45,000 for roadbed construction. These costs vary by location, particularly due to the amount of dump truck travel that is required. Unlike this project, most projects required off-site gravel and off-site disposal of slash and overburden. So, the preliminary estimate for this road was only $25,000/mile for the combined clearing and construction costs. Since construction has slumped some, it seems likely that costs will be 5% to 10% lower than last year. Furthermore, the new route is more direct than the meandering tank trail (2.8 miles vs. 3.5).

The "new" access road would require that a new bridge be built to cross the creek. Because of the importance of this segment, preliminary estimates were already made for three alternatives. The results of the cost estimates for the timber, steel, and reinforced concrete structures are summarized in Table 43-1.

Table 43-1   Bridge Costs (in $1000s)

	Wood	Concrete	Steel	Approaches
First Cost	$350	$575	$480	$300
Painting	$10	$3	$15	
Interval (years)	3	0	5	
Repair	$20	$15	$15	$110
Interval (years)	5	5	5	25
Annual Maintenance	$55	$30	$25	
Life (years)	20	40	30	100

Maintenance for the new road on a per-mile basis will cost about the same as the existing road. These costs have averaged $1500 for grading, $2500 for snow removal, and $1000 for sanding. The only exception is one very steep hill (0.45 miles long) that requires grading twice as often year-round and that incurs 90% of the sanding bill each winter. This hill will be closed to traffic and converted to a sledding hill for the park—if a new access road is built. The rest of the existing road will be maintained for park use only, and this lower level of use should cut its maintenance bill per mile in half.

In evaluating projects of this kind, the city uses a 12% discount rate. Estimate life-cycle costs for the various alternatives, and identify required values for accident prevention and travel-time savings that would support your recommendations.

Options

1. Simplify the problem by only analyzing the bridge alternatives under different assumptions for project life and salvage value, using EUAC and/or PW approaches. Is the assumption of project life important here?

2. If the project does not make sense yet, at what point will traffic increases for the park and the subdivision justify a new access road?

3. Which uncertainties are most critical for project justification: availability of gravel, route selections along park boundary, growth in traffic levels, accident rates, construction costs, discount rate, or ?????

4. Construct and compare three realistic scenarios, ranging from nearly the "most" to nearly the "least" favorable for the new road.

5. Construct a cost/trip measure based on the current level of traffic, on an equivalent annual number of trips, and on the maximum level of traffic expected.

Sunnyside – Up or Not?

Sunny Acres was a desert, then a retirement community, and now it may be becoming a high-tech oasis. If community boosters and the local university can be believed, then Sunny Acres will be to genetic engineering what Silicon Valley and Route 128 are to computers. Sunny Acres may also simply remain a small city with more industry, but still focused on being a retirement community.

The planning staff of the regional public utility commission has described these possibilities as {0%, 3%, or 5%} annual growth with ascribed probabilities of {.2, .5, and .3} respectively. Unfortunately, over the utility commission's standard 30-year study period, the low and high population estimates vary by a factor of 4.3. This variation is a key difficulty for the commission in minimizing the cost of reliable service to the area's population.

One of the questions that the commission must face is how to charge for the cost of capital facilities. If the annual cost is level, then users in early years pay at a higher rate than users in later years. Alternatively, if the commission sets rates based on a level cost per user-year, then "repayment" of capital costs will be very slow initially. In fact, the payments in early years will probably not even be enough to cover the interest on the initial investment.

In this dry area, the largest costs are associated with dams and other water-diversion projects. In addition, these "generation" projects have large up-front costs and fixed capacities; thus, it is for these projects that the uncertainty is of the most concern. Local distribution/collection trunks for sewer, for water, for power, and for electricity are more easily expanded for incremental population increases (tied to new construction). However, a major dam, once the size is chosen and the dam is built, can only be expanded at great cost, if at all.

For example, consider a water-diversion tunnel and aqueduct that are proposed for Sunny Acres. This project will cost $600 million, which will be spread uniformly over a 4-year construction period. When the water project comes on line, there are expected to be 250,000 users in Sunny Acres. Its design is linked to the high growth estimate, so its capacity is adequate for the 30-year horizon. The physical lives of the tunnel and aqueduct should exceed 100 years.

Once it is in operation the annual costs for pumping (including periodic pump replacement and upgrading) will be a function of the growing annual volume. However, these costs can easily be tied to the number of users in each year, so there is little controversy over the correct computational method. These variable costs will simply be added to the capital costs.

Recommend a policy on capital costs for the utility commission, and compare its advantages and disadvantages with other possibilities. Assume that the interest rate is 9% for all calculations.

Option

To simplify the case, students might consider only one pricing alternative and/or one growth rate scenario.

Suggestions to the Student

1. Assume the utility commission uses a level cost per user-year. Now compute the *capital* cost per user under each of the three growth scenarios. Compare the average of these three costs per user-year with a cost based on an average growth rate. Is one value better than the other? Why?

2. Use the cost/user-yr for the average growth rate. Now compute the year of maximum "indebtedness," which is the year when user charges equal interest payments, and the year when 50% of the initial bond amount has been paid off.

3. Assume the utility commission uses serial maturity bonds so that the total paid off each year is constant. This means that the charge/user-year decreases as the

population increases. Compute the capital cost per user-year over the 30 years under each growth scenario.

4. What are the political advantages and disadvantages of each policy? If growth is different than "expected," then what are the risks of each policy?

Case 45

Should the Transmission Intertie Be Built?

Transmission interties are high-voltage power lines that link different electric power generation sources into networks. Goals include increased reliability, decreased power cost, and shared reserve capacity. However, interties cost money to build and operate. Power is lost during long-distance transmission.

In scenic and wilderness areas transmission lines and towers generate controversy. Many consider interties to be visual eyesores which have economic impacts through decreased tourism. Access roads and right-of-ways also increase human pressures on wildlife. In urban and suburban environments there are concerns over a possible link between electro-magnetic force (EMF) and human health. All electricity users receive benefits, but disbenefits affect only those who live, work, or play near the transmission line. This creates political problems due to a NIMBY attitude (not in my backyard).

A proposed 230kV intertie would create a second connection between Anchorage and Fairbanks. This is along a separate geographical route from the existing 138kV transmission line, which was built in the late 1960s. That line is scheduled for major maintenance that will shut it down for 5 construction seasons. The geographical separation means that a single cause, such as an avalanche, cannot sever both links. With this connection, Fairbanks can rely less on coal- and oil-fueled generation and more on cheaper Anchorage sources (natural gas and hydro). With the single existing transmission line the system is vulnerable to failure, and very cold winter temperatures in Fairbanks (it can be -50°F for a month at a time) increase the costs and risks of system failure. The project includes upgrades of stability-enhancing equipment, which would also improve performance of the existing intertie within the Matanuska Valley (en route from Anchorage to Fairbanks). The stability-enhancing equipment (SEE) could be installed by itself, but the intertie requires the SEE.

Capital costs for the stability enhancing equipment total $20.5M. In addition, $50,000 per year will be spent on insurance and maintenance for it. The equipment will have no salvage value after 25 years. To simplify the analysis, assume that the costs for the equipment and the intertie occur at time 0. This is basically equivalent to treating the project start-up as time 0 with the future worth of the investment cost incurred at that time.

The new 230 kV intertie costs $57.1M. The towers would be steel. Most utilities assume a 50-year life; however, most steel-pole lines that have been constructed are still operating. This includes a 1932 project by B.C. Hydro and a 1904/06 project by Pacific Gas & Electric. Expected costs to maintain and operate the line average $145,000 annually, stated in constant value dollars.

The intertie would increase system reliability in Fairbanks, and the stability-enhancing equipment would enhance reliability in the Matanuska Valley. The value to the utilities of increased reliability is based on saving 12% of the Fairbanks and 10% of the Matanuska Valley unserved energy (= amount not received due to outages). Economic values for this unserved energy equal the cost of being without power. Recent estimates from different studies are $2.18, $3.45, and $9.78/kWh for residential customers. At $5/kWh, the outage cost to Fairbanks residents is $.18M/year. For Matanuska residential customers the cost is $.68M/year. For commercial and industrial customers much of the cost of power outages is a fixed cost if the outage occurs. Thus, a 20-minute outage is only 3 times as expensive as a 1-minute outage. For these customers the outage cost is $2.71M/year in Fairbanks and $1.87M/year in the Matanuska Valley.

The project's principal cost justification is to substitute Anchorage's lower power generation cost of $18.05/MWh for Fairbanks' higher cost of $33.47/MWh. Power transfers from Anchorage to Fairbanks would increase by 67 GWh/year by adding the stabilization equipment—measured in Anchorage before transmission losses. Over 15 years this would increase to 106 GWh/year, where it would remain for the rest of the project's horizon. The transmission loss is 33% for the current transmission link. If the stabilization equipment and the new intertie are added, power transfers would increase by 122 Gwh/year, which would increase to 186 Gwh/year over 15 years. The higher capacity intertie dramatically reduces transmission losses. At 122 GWh/year losses would be only 2 GWh/year, and at 186 they are only 23 Gwh/year.

The contractual arrangements between the Anchorage and Fairbanks utilities are still subject to negotiation, but the perspective of the state power authority is clear. What is the bottom line for the state? From that perspective the value of the transferred power equals the power received at the Fairbanks price minus the cost of the power generated at the Anchorage price.

The forecasting of future electric power needs is difficult, as can be illustrated by the Washington Public Power Supply System (WPPSS) or "whoops." To meet anticipated increased power needs, five new nuclear power plants were once in progress. When demand failed to materialize, only 1 was completed, 1 was left incomplete, and 1 was taken back down to the ground. In 1983, WPPSS defaulted on $2.5B on bonds issued for the construction of the facilities.

An intertie allows generation systems to share reserve capacity. This computation is complex, since it includes some capacity expansions that can be deferred and other expansions that can be avoided. The result can be approximated as $455,000/year for the stabilization equipment and another $420,000/year for adding the new 230kV intertie.

The final benefit from the new 230kV intertie is that reconstruction of the existing line can be delayed until the new line is complete. Then there will not be power interruptions during 5 years of 7-month construction seasons. The cost of using more expensive Fairbanks-generated power would be $7.7M/year for each construction season. If the intertie is not built, those costs will start next year.

The power authority uses a 4.5% real discount rate (all costs and benefits in constant value dollars). There will be a 50/50 cost sharing between the utilities and the state power authority on all first costs. What should the state power authority recommend? How did you value the disbenefits to scenic and wildlife values? Is there a difference in the state's and the utilities' conclusions on the wisdom of these projects?

Prepare an analysis to support a perspective—objective, green, or pro-development.

Case 46

Aero Tech

Aero Tech's executive committee has spent the last two weeks debating the annual capital budget. They are nearly ready to make a decision, which they would normally do in their Tuesday meeting. Ms. Gloria Burns is the CEO, and she has asked them instead to write memos. Her intent is to use these memos to help educate two new members of the Board of Directors: you and Mr. Fred Vail. At the same time, she can use your reactions as a double check on the approach that has historically been used.

The committee has agreed that the total operating budget for next year should be $124 million. This leaves $24 million for capital investment, although additional funds could be raised through borrowing or through the issuance of new common stock. Ms. Burns has decided that the stock possibility should be ignored; in fact, the executive committee is even considering using part of the capital budge to *buy back* some of the company's stock. The finance VP estimates that this will have a 15.4% rate of return, as long as no more than $5 million is repurchased. This repurchase would be part of the capital budget.

If bonds were sold to increase the $24 million, then the cost would be 10.8%, with semiannual interest and a 10-year life. These bonds could be for any amount between $1 million and $5 million, and the setup fee would be 2% of the amount borrowed. If a loan were used, the rate would be 12.4%, but *all* payments would be made at the end of the loan. The period of the loan could be anywhere from one to three years. The amount borrowed could be as low as $500,000, but the upper limit is only $2 million. In this case, the setup fee is 1% of the amount borrowed. The total increase in debt is limited to $6 million.

Historically, Aero Tech has used a 10% minimum attractive rate of return in evaluating its capital projects, so that projects whose present worth was negative were rejected. They have not standardized on an approach to prioritizing those projects that meet the 10% minimum criterion. Usually they fund from one-third to two-thirds of the projects that meet the minimum criterion.

Develop a recommended capital project list and budget for Ms. Burns, including a description of your recommended methodology. This memo should also include a critique of the proposals from members of the executive committee. The candidate projects are listed in Table 46-1. These projects are independent of each other, and any combination can be accepted subject to the funding limits.

Table 46-1   Proposed Projects for Aero Tech

Project	First Cost ($ \times 10^6$)	Net Annual Benefit ($ \times 10^3$)	Life (years)
A	5.5	1420	10
B	8.1	2500	5
C	4.7	700	15
D	3.9	830	25
E	2.1	400	15
F	5.0	970	20
G	6.3	1310	10
H	1.9	730	5
I	2.4	460	10

To: Gloria, CEO
From: George, VP for Engineering

The table below shows the projects that I recommend. Notice that this depends on bonding $4.8 million for five years, so that we can add project B to the funded list. The cutoff is highlighted.

Table 46-2   Engineering's Proposed Projects for Aero Tech

Project	First Cost ($ \times 10^6$)	Net Annual Benefit ($ \times 10^3$)	Life (years)	Present Worth at 10% ($ \times 10^6$)	Cumulative First Cost ($ \times 10^6$)
D	3.9	830	25	3.634	3.9
F	5.0	970	20	3.253	8.9
A	5.5	1420	10	3.225	14.4
G	6.3	1310	10	1.749	20.7
B	8.1	2500	5	1.377	28.8
E	2.1	400	15	.942	30.9
H	1.9	730	5	.867	32.8
C	4.7	700	15	.624	37.5
I	2.4	460	10	.427	39.9

To: Gloria, CEO

From: Simon, VP for Manufacturing

The table below shows the projects that I recommend based on a payback criterion. Notice that this depends on borrowing $1.7 million for at least a year. I prefer this to the bonds because of the lower setup fees, and our ability to pay it off at our convenience. The cutoff is highlighted.

Table 46-3	Manufacturing's Proposed Projects for Aero Tech				
Project	First Cost ($ × 10^6)	Net Annual Benefit ($ × 10^3)	Life (years)	Payback Period (years)	Cumulative First Cost ($ × 10^6)
H	1.9	730	5	2.60	1.9
B	8.1	2500	5	3.24	10.0
A	5.5	1420	10	3.87	15.5
D	3.9	830	25	4.70	19.4
G	6.3	1310	10	4.81	25.7
F	5.0	970	20	5.15	30.7
I	2.4	460	10	5.22	33.1
E	2.1	400	15	5.25	35.2
C	4.1	700	15	6.71	39.9

To: Gloria, CEO
From: Aristotle, VP for Finance

The table below shows that I have evaluated the projects at a discount rate of 18.4%. This is the average rate of return for the projects that we accepted last year. Since some projects are necessarily below average, this means that at least one of my recommended projects is shown with a negative present worth.

I am comfortable with the top 4 projects. The next 3 projects (E, B, & I) are worthwhile, but I'm not sure which option is best.

1. Do E and buy back $1.6 million in stock.
2. Do B and borrow $500,000 ($100,000 put in bank).
3. Do E & B and borrow $2.5 million through bonds.
4. Do E, B, & I and bond for $4.9 million.

Table 46-4   Finance's Proposed Projects for Aero Tech

Project	First Cost ($ × 10⁶)	Net Annual Benefit ($ × 10³)	Life (years)	PW at 18.4% ($ × 10⁶)	Cumulative First Cost ($ × 10⁶)	
A	5.5	1420	10	792	5.5	
D	3.9	830	25	545	9.4	
H	1.9	730	5	362	11.3	
F	5.0	970	20	92	16.3	
E	2.1	400	15	-99	18.4	cutoff
B	8.1	2500	5	-352	26.5	???
I	2.4	460	10	-362	28.9	cutoff
G	6.3	1310	10	-495	35.2	
C	4.7	700	15	-1198	39.9	

Suggestions to the Students

1. One of the difficulties in comparisons of these projects is that they do not all have the same life. Does present worth properly adjust for this? Does payback period?

2. If not, can modified measures be constructed? If not, what criteria should be used for ranking?

3. One of the decisions you may have to make (depends on the criteria you use) is the appropriate discount rate for evaluation of these projects. Two considerations in this choice are (1) the opportunity costs of rejected projects, and (2) the reinvestment rate you expect to apply to future choices.

Bigstate Highway Department

Bigstate's highway department must select projects to be built during the coming year. Eleven projects have passed the preliminary screening (see Table 47-1). Five projects have mutually exclusive alternatives (for example A1, A2, and A3 are mutually exclusive alternatives for Project A). Each project and its alternatives must be evaluated at a 10% discount rate. The best set must be selected within a budget constraint of $36 million for the first year's construction cost. For this analysis all projects and alternatives can be assumed to have a life of 20 years.

Political considerations require that at most one project of A and C be selected. Similar considerations require that at least three projects be selected from projects A, B, D, F, H, and J.

One of the biggest problems you have is that the Commissioner has identified the value for saving a human life with a range of $100,000 to $500,000, rather than a single value. Similarly, the department has determined that preventing a serious injury has a value between one-quarter and two-thirds of the value of a saved life.[1]

What projects do you recommend?

---

[1] Liability lawsuits often result in higher awards for serious permanent disabilities than for deaths. The basis is not the "value" of the injury, but the cost of care and compensatory damages.

Table 47-1	Bigstate's Possible Highway Projects				

	First Cost ($M)	Annual Operating Cost ($1000s)	Annual Benefits		
			Time and Distance ($1000s)	Lives	Injuries
A1	8.0	350	950	2	11
A2	10.5	550	1300	2	11
A3	14.2	950	1350	2	11
B1	12.0	400	500	4	8
B2	14.5	500	1600	6	9
C	4.3	150	1100	0	0
D1	7.5	500	1950	1	8
D2	11.0	800	1950	4	10
E	9.4	0	700	3	5
F1	6.6	100	1100	0	6
F2	9.3	150	1500	0	8
G	7.5	200	1150	0	0
H	5.6	300	1500	1	0
I	7.9	1500	2800	1	1
J1	8.3	275	950	0	3
J2	11.4	475	950	2	3
K	16.0	850	2500	1	8

Option

How do your recommendations change over the range of possibilities for the values of lives and injuries? This can be presented graphically or in statements similar to the following: "For values of human life between $175,000 and $238,000, do projects A3, D1. . . ."

Dot Puff Project Selection

The Department of Transportation, Public Utilities, and Facilities is affectionately known as DOTPUF. Its broad responsibilities include design and construction of public highways, office buildings, and other state projects. This does not include selecting the projects. Instead, selection is part of the state's budget process, and it is "managed" by the governor's office and the legislature.

DOTPUF's involvement with a project begins with a request from a city or regional government, a legislator, a citizen, or some public interest group. A *preliminary screening* compares the project's benefits with its costs at the legislatively mandated 10% discount rate. If it passes, then DOTPUF conducts an in-depth *feasibility study*. At this stage some projects are eliminated for failure to "pay off" at the 10% discount rate. For the other acceptable projects, a benefit/cost ratio is calculated to quantify project desirability. These ratios certainly influence the selection decision, but they are not the only factor. The cost estimates of the feasibility studies are also used in budgeting for the selected projects. Once the selection has been made, DOTPUF must write all specifications, oversee the bidding process, and contract with the engineering and construction companies.

Not surprisingly, the selection process sometimes emphasizes the power of particular legislators or regions that strongly support the incumbent governor. As a result, some unfunded projects have "higher value" than others that get built. For example, consider last year's projects in the Northcentral Region. The ten accepted projects had benefit/cost ratios ranging from 1.21 to 1.78, while the ratios of the eight rejected projects ranged from 1.03 to 1.32. Of the 10 projects that were funded, 4 ranked lower than the project with a 1.32 ratio, which was rejected. Three of the eight rejected projects ranked higher than the project with a 1.21 ratio, which was accepted.

The selection process considers the projects' capital cost, but not their demands for DOTPUF resources. Because DOTPUF contracts out most of the detailed engineering and all of the construction, the variety of project types and sizes leads to a poor correlation between DOTPUF's time commitment on a project and its total size. In fact, the smallest jobs are sometimes designed in-house, so they may have the *largest* time commitment. Table 48-1 summarizes the unevenness in workload that results, although the budget process has enforced some regularity on the total construction cost to the state.

Table 48-1	Historical DOTPUF Project Summary			
Fiscal Year	2006	2007	2008[a]	2009
Number of projects				
Submitted	153	205	187	172
Accepted	128	150	148	135
Funded	72	80	44	65
Capital cost ($\times 10^6$)				
Accepted	240	290	325	250
Funded	130	140	185	140
As built	140	135	210	125
DOTPUF man-years[b]				
Funded	31	55	72	28
As built	37	52	60	34

[a] 2008 had a very large contract for a new state office building, which required lots of coordination in planning. More problems with projects occur when DOTPUF is overloaded (2006 overrun).
[b] DOTPUF's staffing level is stable at 50, so man-years expended tend to move toward this level from the desirable level identified during the feasibility study.

Floyd Ackroyd, the deputy administrator, is the ranking civil servant (permanent employee) for the agency. His boss, Conrad Couch (better known as Conrad the Couch-Pumpkin), has started job hunting, since the replacement for the lame-duck governor is certain to replace him as well. Floyd wants to improve the image and the performance of DOTPUF, and he is convinced that the time is right for attempting to change the historical process. He believes that the leading gubernatorial candidate, who is an engineer, will be receptive to a well-supported argument for change. Floyd also knows that virtually all of the agency's employees are dissatisfied with the results of the current process. In years when the workload is high, everyone is pressured to work overtime—without additional compensation. In years when the workload is low, many are frustrated by the need to appear busy.

Floyd is satisfied with the performance of his department in all stages of this process—he sees the main problem as unreasonable expectations from external constituencies. There are two manifestations of this problem. Luckily, they are not misestimates and cost overruns. Rather, they are the dissatisfaction of citizens who have waited years for an obviously worthwhile project and the frustration of everyone when the workload is too high and DOTPUF falls behind schedule. Outsiders decry the inefficiency that produces the delay, but Floyd attributes it solely to the difficulty of a workload that fluctuates by a factor of 3.

Floyd believes that a possible solution lies in his department doing more—not less. He believes that the department should attempt to guide the selection process, rather than simply analyzing project feasibility and waiting for the results of the selection.

When he discussed this with Conrad, the chief disagreement was over the best measure of quality. Floyd is comfortable with using the same benefit/cost ratios that have been used in the past. Conrad suggests that is politically necessary to change the measurement somehow. Perhaps to calculate the benefit/cost ratios at a different discount rate, or perhaps to rank on the rate of return "earned" by each project. Conrad also disparaged the usefulness of a priority ranking on the projects. Instead, he suggested blackballing projects and removing them from the list.

Conrad told Floyd to go ahead, but that he would not participate himself. With the go-ahead Floyd had decided to proceed with a trial run. Specifically, he has asked you to analyze the project for the northcentral region of the state (see Table 48-2).

To quote from his comments at the recent meeting of his department heads: "I believe that a large fraction of our perceived performance problems stem from approving too many projects. As a result, we are not controlling which projects are selected for funding. I do not believe that we can select the projects, but we should be able to rank them in priority and perhaps identify some projects as *not recommended for funding*. Some of these projects will meet our first-cut criteria, and this will be a second cut. I have asked the northcentral regional office to be the guinea pig. Then, once they are done, we will discuss how we might be able to do better, so that every unit can complete the same exercise."

In his conversation with you, he relayed Conrad's comments and indicated that the northcentral region was selected because of similarities between this and last year. These similarities include project mix and project "quality" for the feasibility stage list, and they also include probable funding and staffing to match last year ($20M and 8 man-years). Besides the priority listing and the blackballing of some projects, he needs a comparison of the probable results using the old and new processes. As an aside, he explained that he is not using historical data and old lists of potential projects, because of the political problems attendant on picking out projects as "mistakes." He stressed that it may be more difficult to respond to large fluctuations in workload than to achieve political support for a certain level of funding.

Table 48-2	Proposed Projects for Northcentral				
Project	First Cost ($ $\times$ 10^6)	Net Annual Benefit ($ $\times$ 10^3)	Life (years)	Required Man-Years	BC at 10%	IRR (%)
A	1.2	270	10	.8	1.38	18.31
B	2.6	415	20	1.1	1.36	14.98
C	.8	140	20	.6	1.49	16.70
D	5.2	1075	10	1.5	1.27	15.98
E	.5	90	10	.4	1.11	12.41
F	.2	30	20	.1	1.28	13.89
G	1.9	250	40	1.8	1.29	13.06
H	4.5	910	10	1.9	1.24	15.39
I	3.9	845	20	1.2	1.84	21.20
J	2.7	600	10	1.1	1.37	17.96
K	1.1	230	10	1.1	1.28	16.28
L	.6	90	20	.4	1.28	13.89
M	.4	51	20	.5	1.09	11.23
N	2.4	660	10	1.0	1.69	24.40
O	3.1	470	40	2.1	1.48	15.11
P	.6	125	10	.8	1.28	16.18

Case 49

The Arbitrator

AccuSpeaker manufactures stand-alone speakers including the enclosure, which are then sold under the labels of AccuSpeaker at major retailers. The speakers are also sold without enclosures to other manufacturers. Your new job is as staff assistant to Henry Higgins, the CEO. One of your first tasks is to evaluate the proposed capital budgets of the manufacturing and marketing department heads. These proposals represent the second step in the annual budgeting exercise, which begins with the CEO and the department heads identifying the projects that might be undertaken. After each of the four department heads ranks the alternatives, the CEO evaluates each proposed budget. He then prepares a budget that is the basis for further discussion, modification, and finally adoption.

Henry Higgins is a firm believer in training and testing. Evaluating all four proposals would probably be overwhelming, so he only gave you the memos from two departments. He has also decided *not* to guide your thinking by letting you examine last year's proposals and summation. By forcing you to compare two budgets, the CEO can observe your skill at ferreting out the truth from alternative perspectives. By examining your methodology, he can check for other theoretical approaches that may differ from the firm's historical techniques.

Henry has identified $1.5 million as next year's capital budget, although transfers between the capital and operating budgets gives some flexibility. He has indicated that the firm is borrowing money with long-term loans at 9%, that last year's returns on sales was 6%, and that contribution to profit and overhead is about 65% for most products.

To: Executive Committee
From: Henry, CEO
Date: October 17
About: Minutes of Meeting on Potential Capital Projects

We discussed numerous projects at our recent meeting, but I believe that we reached a consensus that the following three (with some mutually exclusive alternatives) were the ones worth further consideration. We also agreed to follow the same process as last year. The next step is to prioritize the projects with their preferred alternatives. Please have your memos on your proposed capital budget to me by October 22, so that we can refine these before the Board of Directors' meeting in early November.

Project I: Installation of CAD/CAM System

The engineering department has been using computers for analysis of proposed designs for years, but we have not yet attempted to integrate this with the drafting or manufacturing processes. They have suggested that we purchase an integrated system from the ADAM Corp. (Aiding Design And Manufacturing). The proposal can be undertaken immediately, or we can do a part of it now and decide later on the rest. We believe that ADAM will cut the price for each module by 10% for each one we buy at the same time. Thus if we buy alternative 3 immediately, then the prices for the analytic, the drafting, and the control modules will all be cut by 20% (both hardware and software). If we wait to purchase the more advanced modules, some decline in hardware costs is likely. They charge a 15% annual fee for maintenance and updates.

Alternative 1: Replace our current computer system with ADAM's analytic module. This would provide the foundation for purchase of other modules later, and it would allow us to better judge ADAM before we commit substantial sums. This provides little in new capabilities, but would require about 6 man-months in training for users (engineers) at $5000 per month including benefits. The cost for hardware is $50,000 with a 10-year physical life, while the software will cost another $50,000.

Alternative 2: Add the drafting module. This would cost $75,000 for software and $15,000 for hardware (life about 5 years). This would allow us to eliminate two drafting positions, although about 3 man-months of extra time will be required from one of the engineers.

208

Because the computer cannot be as flexible as a person, we will have to standardize some elements of our design.

Alternative 3: Can be done with or without 2. Add computerized control of some of the machining and cutting operations. While the additional cost for computer hardware is inconsequential, this would require about $350,000 in new factory equipment and process control hardware. This new equipment should last 10 years. About three-quarters of this investment is likely to be required within the next 5 years anyway, when our current equipment will have to be replaced. The other quarter is the cost of controlling hardware. This is a standard application, so the software would only cost $50,000, and we believe we could reduce the number of machinists from 4 to 2.

Alternative 4: Add computerized assembly operations. After our meeting I decided to remove this alternative from current consideration. I cannot yet accept the large cost and uncertainty in moving to the use of robots. I am sure the Board would seriously consider such a proposal, but only after we have implemented at least the drafting or machining modules.

Projects II & III: Adding Cabinetry Facilities

Currently we subcontract for both our wood and our plastic speaker enclosures. By purchasing appropriate equipment we could move one or both types of enclosures in house. Either one would be sufficient to allow greater flexibility in our production process. To have more flexible responses to machinery and supply problems is worth about $80,000 a year to us.

Alternative 5: Add woodworking facilities. We currently purchase 55,500 wood enclosures each year at an average cost of $8.57 each. Manufacturing's estimate is that these could be produced for $5.00, including costs of material, the operating costs of the machinery, and the labor and benefits of the new employees. The new machinery would cost about $200,000 and would last 15 years. The storage of incoming materials would save a small amount of space, because we would have a smaller inventory of completed enclosures, and the machinery for "finishing" would fit in an unused portion of the plant (1500 square feet). However, the cutting, sanding, etc. equipment would require a new 2500-square-foot addition, which would cost $175,000 and last at least 40 years.

Alternative 6: Add a small injection molding facility. We now purchase 45,000 plastic enclosures each year at $3.58 each. This has essentially the same space requirements as the woodworking option, but the capital cost is double that of the woodworking alternative for any volume between 30,000 and 100,000 per year. The first shift can run at 30,000 to 50,000 units, while the second shift can be part-time. A major refurbishing would be required about every half a million units for $250,000. Because of the automation built into this continuous process, the combined cost for material, for energy, and for operator labor is only $0.35 each (0.38 on the night shift).

Alternative 7: Add a larger injection molding facility for plastic enclosures. The space requirements would increase by 25%, the capital costs by 50%, and the capacity by 100%. The unit costs would remain about the same, but the refurbishing interval would double with the capacity.

Project IV: Revamp Electronics Assembly Line

This is the only portion of our operation where we suffer from significant worker dissatisfaction. We have reorganized our other assembly lines with two main operational criteria—increasing the number of tasks performed by each employee and physically placing them so that they can converse while they work. The results over the six lines have increased our productivity by 15%, 24%, 31%, 10%, 18%, and 5% in order of introduction. While this line will be more expensive to change than our last one, there is far more potential for improvement. However, I suspect that some workers prefer the older linear assembly line, and that they are shifting positions with other employees who prefer the newer, more team-like organization.

Alternative 8: Modernize the equipment without restructuring the line. This would keep one line in the traditional mode. Since we have modernized our other lines, this is currently our "oldest" line. It has the largest incidence of equipment problems. At an installed cost of $250,000 for new equipment, we can reduce our annual maintenance costs by $35,000. More importantly, productivity should increase by 5% without any increase in operating costs. The line produces nearly 800,000 units per year with a direct manufacturing cost of $5.40 (excluding materials). Although this revamping will extend the life of the line for only ten years, it is unlikely that we can put it off for more than 2 years.

Alternative 9: Rebuild and restructure this line. This would increase the capital cost by 50% over the modernization alternative. It will improve productivity by the same 5% as modernizing the line by eliminating the reliability problems. It may produce large productivity gains due to the reorganization. We have experienced large productivity gains on all of the other lines.

To: Henry, CEO
From: Myron, Head Marketing
Date: October 17

I recommend that we accept alternatives 3 and 7. While I have calculated the present worth indexes,[1] I find that I place more importance on two other reasons. I support alternative 3 (and 4 likewise) because I believe it will allow us to respond more quickly to customers who want to include our speakers in their systems. I cannot quantify the value of this, but I believe that it is essential for our competitiveness. Secondly, I support alternative 7, because I believe that we can substantially increase the sale of our speakers in plastic enclosures. If one examines the national market shares of wood and plastic enclosures, there is a steady trend of 2% annual shift to plastic.

---

[1] These have been omitted to provide a student option. The present worth indexes should be calculated as the PW of the annual net receipts (including costs) at 6% (our rate of return on sales) divided by the PW of capital expenditures. Use the time period of each alternative and do not analyze the increments between the alternatives within a project. Include direct labor and benefits only.

To: Henry, CEO
From: Margaret, Head of Manufacturing
Date: October 21
About: Next Year's Capital Budget

I recommend that we accept alternatives 2 and 9, which would allow the purchase of ADAM's computer aided design and drafting modules and reorganize our production line. I analyzed the alternatives within each project to decide which one was best, although the enclosure project would allow alternative 5 along with one of 6 or 7. Even though both projects 5 and 6 appear to be good choices, and even though both can be done within the capital constraint, I would prefer to reserve that capital for the implementation of alternative 4—which is likely to be quite expensive.

Thus my recommendations depend more on the urgency of certain improvement than on the payback periods that I calculated.[2] We know we are going to redo that assembly line— let's get it done. Then it won't be hanging over our heads.

Also I believe we must implement a CAD/CAM system, but I've heard too many horror stories when the process is rushed. While alternative 3 is attractive, I believe that we would get more benefit out of computerizing our assembly operations than adding new numerically controlled machines. Because you feel the Board won't accept alternative 4 yet, I think we have to build a base for selling it. Alternative 1 will have great difficulty in implementation, because it is all cost and no benefit—thus only alternative 2 is left.

I realize there is some conflict between my desires to begin computerizing assembly operations in the near future and to immediately restructure the electronics line. However, I believe that we cannot maintain a line on straight hourly wages with a production bonus unless the line functions reliably. This is particularly true since everyone in the plant knows how well some of the small production teams are doing in incentive bonuses, not to mention the posters for team of the week (which in four years has never featured a team from our older lines—every winner has been a team from a reorganized line!!!)

---

[2] Calculations of payback period should include a burden rate of 120% for indirect expenses and overhead, which should be calculated on direct labor hours and benefits. The base wage is $9.87/hour.

Case 50

Capital Planning Consultants

Your consulting firm has been hired to assist WWTWDW (We Wish This Were a Deterministic World) with their capital budgeting. You are to write a brief "technical report" giving the recommended projects, and describing your principal assumptions. What is your evaluation criterion, and why did you make this choice? As some estimates may change before project selection is actually made, discuss which circumstances would force changes in the recommended list and what those changes would be. The analysis used in developing the recommendations should be an appendix to your technical report.

1. The four numbered projects (Table 50-1) may all be done, but the lettered alternatives within each project are mutually exclusive.
2. The minimum attractive rate of return is 15%, but the firm averages a 20% return on its investments.
3. The capital budget, excluding borrowing, is $100,000, and the company may borrow on short notice $10,000 at 18% from an interested venture capitalist. (The borrowing rate includes an increase that reflects the Board of Directors' estimate of the cost of increases in bookkeeping expenses and managerial time used.)
4. You may assume that pessimistic estimates have a probability of .25 and that optimistic estimates have a probability of .1.

Table 50-1	**WWTWDW Capital Projects**			
Project Alternative	First Cost	Annual Benefit	Useful Life	Salvage Value
1A	20	4.9	(8, 10, 12)	0
1B	30	7.5	(8, 10, 12)	0
1C	35	9.1	(8, 10, 12)	0
1D	40	10	(8, 10, 12)	0
2E	24	(4, 6.5, 7.5)	(6, 10, 15)	(0, 0, 5)
2F	25	(2, 8, 9)	(8, 10, 12)	(0, 0, 10)
3G	20	(4, 5, 6)	(8, 10, 15)	0
3H	35	(9, 12, 18)	(8, 10, 15)	0
3I	(40, 50, 60)	(15, 20, 30)	(8, 10, 15)	(0, 0, 10)
4J	15	(3.5, 4, 4.5)	(8, 9, 12)	0
4K	(20, 25, 30)	(3.5, 7, 10)	(7, 10, 15)	0

Note: All monies are in $1000s.

Case 51

Refrigerator Magnets Company

The Refrigerator Magnets Company (RMC) makes magnets. The magnets are sold in a wide range of consumer outlets around the country. The annual capital budget is due, and as the engineering manager you have been asked to review the proposed projects submitted by the various departments and to prepare a final list for board approval. The board has budgeted $1,200,000 for capital improvements. Last year's minimum attractive rate of return (MARR) was 12.8%. The weighted average cost of capital on which the $1.2 million capital budget was based is 8.5%, and the marginal cost of capital is 11.5%.

The projects listed in Table 51-1 have been submitted. Regulatory projects are obligatory to meet state or federal requirements. Product growth projects are needed to support increasing product demand, and the project's benefits are from increased sales. The operating budget for next year assumes that the additional capacity will be created in a timely manner. Cost reduction projects are just that—projects which will reduce operating costs. They are discretionary. The new-product projects are to allow additional items to be added to the company's product offerings and as such increase sales (and thus profit).

Projects B and C are mutually exclusive solutions to the same problem. Projects L and M are dependent on Project K, but Project K does not depend on L and M. The production manager has been pushing to combine these three projects into one project on the basis of the synergy to be gained from three related parts on a common process.

Capital projects at RMC have historically been completed at 105% of the budgeted cost (on average). While company policy allows a +/-10% deviation (or a maximum of $10,000) on a project's capital expenditures without major questions or a supplemental capital request being required, the board has expressed its "strong desire" that the capital budget not be exceeded. From past experience you know that several small projects will come up during the year, which will be required for unforeseen opportunities or problems. You believe that $50,000 should be set aside for these projects.

If a project is listed on the approved capital budget, project approval upon submittal of the detailed capital appropriation request is pretty automatic. When projects not on the capital budget are submitted (either as substitutions for budget projects or as opportunistic projects), they receive closer scrutiny during the evaluation/approval/disapproval process.

What is your final project list for the capital budget? What is your MARR recommendation for the upcoming year? Assume that all expenditures will be made in the budget year.

Table 51-1   Project List

Project	Type	First Cost	Life	Salvage Value	Annual Benefits
A. Packaging Line Automation	Cost Reduction	$150,000	7	$20,000	$32,000
B. EPA Compliance – Option 1	Reg. Reqt.	75,000	10	0	0
C. EPA Compliance – Option 2	Reg. Reqt.	140,000	10	20,000	18,500
D. Automated Stamping Press	Cost Reduction	185,000	8	45,000	38,000
E. Banbury Mixer Upgrades	Cost Reduction	200,000	6	0	48,000
F. Glue System Upgrade	Cost Reduction	80,000	3	0	37,000
G. Curing Oven Expansion	Product Growth	129,000	12	22,000	23,500
H. Waste Water Treatment Project	Reg. Reqt.	205,000	15	0	23,000
I. Warehouse Automation – Phase 2	Cost Reduction	275,000	10	80,000	47,000
J. Part # 882G Die Replacement	Cost Reduction	29,000	4	1,000	9,700
K. Transfer Molding Press – New Product	New Product	130,000	6	30,000	28,900
L. Part # 638 Die – New Part	New Product	30,000	6	1,000	9,100
M. Part # A28C Die – New Part	New Product	24,000	6	1,000	5,500
Total		$1,652,000			

Case 52

Aunt Allee's Jams and Jellies

Coauthored by
Neal Lewis
University of Bridgeport

Aunt Allee's Jams and Jellies was founded by Allee Glover and her husband Mangrum in the kitchen of their modest farm in 1958. The original products that were produced by the company were cherry jam and cherry jelly, using an old family recipe and fruit grown on the family farm. As demand grew, cherries were bought from local farms (all within 50 miles of the factory).

By 2008 the business had grown to occupy a modest 20,000-square-foot factory/office building on the site of the old farm. In 2008 the company was still family owned and was operated by four of the founder's grandchildren. Hobb Glover is the company's president. He manages the business and generally helps out in the office as needed. His sister, Amanda Smith, does the marketing and selling of the products. Their cousin, Conrad Glover, is the business's accountant and materials manager. Another cousin, Edith Woods, is the production and maintenance and engineering manager.

By 2001 demand for the cherry jam and jelly had outstripped the company's ability to produce enough jam and jelly from fresh fruit in season. Production was converted to using concentrates and frozen cherries—but all from locally grown fruit. In 2006 the current plant was built with the idea that the company would diversify its product line by introducing two new jams, white grape and blueberry, and a new jelly, white grape, using other fruits available in the state. Test marketing had shown all three products to be popular, and each could command a premium price with higher markups than the current cherry products.

The new products were introduced in late 2006 and were in full production throughout 2007. The family was pleased that the new jams and jellies performed as well as forecast.

They were also pleased that the sales of the cherry jelly and jam were not depressed by the sales of the new products. Unfortunately, the financial results (Table 52-1) were disappointing since traditionally the company has achieved an average of 43% return on sales.

Table 52-1   Financial Results

	Cherry		Grape		Blueberry	
	Jelly	Jam	Jelly	Jam	Jam	Total
Sales	$576,000	$432,000	$145,455	$72,730	$36,669	$1,262,854
Materials	230,400	172,800	43,637	21,819	12,101	480,756
Direct Labor	30,789	23,092	16,986	8,493	4,057	83,418
Overhead @ 13.6% of sales	78,506	58,879	19,825	9,913	4,998	172,120
Tot. Oper. Income	236,305	177,229	65,008	32,505	15,513	526,560
Return on Sales	41.0%	41.0%	44.7%	44.7%	42.3%	41.7%

The factory is operating near capacity, so some decisions need to be made on next year's production. In January 2008, Hobb calls a meeting with Amanda, Conrad, and Edith to discuss the financial results and decide on what needs to be done.

Hobb opens the meeting with a review of the company's current financial situation. While return on sales are only slightly lower than before plant expansion, the increased profit from the new products was being counted on to pay the loan on the new building. Hobb and Conrad do not believe that the company can afford to expand again for at least three more years.

Amanda reports that she has been looking at extensions to the product line. The current products are meeting demand, and she does not believe that demand will increase substantially. She has identified two additional jams which can be sold in quantities similar to blueberry and which should return similar margins. She recommends cutting back on cherry jelly production, if necessary, in order to fill all the orders she can generate for the higher margin items.

Edith reports that the new products cause several problems in production. The main problem is in the setup of the production run. In the past, setups were relatively painless.

Since both products were the same color and flavor, changing over to run a different product was just a matter of emptying and sanitizing the lines. With the current products, when flavors are changed, an extensive changeover is required. Where in the past setups took less than 30 minutes, they now take more than an hour, depending on the product to be produced, because some changeovers require more cleaning and sanitizing between products. She has had to increase hours worked per day to 10 to offset the setups. Edith has so far kept the workweek to four days with no overtime but warns that the limit has been reached with the current crew size. Another shift will be necessary if demand increases.

Conrad reports similar problems in the production control and the accounts payable functions. He reminds the management team that they have had to add one person to the office to support the additional support work required by the diversification in the product line. The additional workload is being caused by more items needing to be tracked, such as labels, fruit, and additional ingredients; more payments to be processed to the fruit growers; and more invoices to check. While Conrad is happy with the sales and apparent markups achieved by the new products, he openly wonders if the current costing system (where overhead allocations are based on dollar sales) is misrepresenting the true costs and thus the true returns on the new items. Conrad suggests reanalyzing the costs based on activity-based costing.

The management team agrees that it is difficult to make decisions if there is concern about the cost system, especially from the head of accounting. Conrad and Edith are tasked with looking into what the five products are really costing Aunt Allee's Jams and Jellies.

As a first step, Conrad and Edith tabulate production requirements and production rates (see Table 52-2). Edith notes that the batch size differences are caused by the size of the different kettles used for each product. Edith also notes that the reason for the different setup times is the difference in the ease of doing the clean-out required between batches. During setup, three people of the four-person crew work on setup while the fourth attends to preventive maintenance and other indirect labor activities. During processing, four members of the crew are needed.

Table 52-2	**Production Data**				

	Cherry		Grape		Blueberry
	Jelly	Jam	Jelly	Jam	Jam
Units produced	128,000	96,000	29,091	14,546	6,667
Price per unit	$4.50	4.50	$5.00	$5.00	$5.50
Batch size, gallons	200	200	50	50	50
Batches produced	42	32	38	19	9
Direct material/unit	$1.80	$1.80	$1.50	$1.50	$1.82
Direct labor hrs/batch	58.5	58.5	35.5	35.5	37.0
Direct labor hrs/100 units	1.92	1.92	4.67	4.67	4.87
Direct labor hours	2463	1847	1359	679	325
Direct labor $	$30,789	$23,092	$16,986	$8,493	$4,057
Units/batch @ 95% yield	3040	3040	760	760	760
Setup time/batch	0.5	0.5	1.5	1.5	1
Administration	1	1	1	1	1.25

As a second step, Conrad and Edith look at the overhead dollars. Conrad notes that the overhead rate was just over 11% of sales two years earlier. The overhead is broken down into five categories (see Table 52-3). The remaining expenses (the salaries of the cousins, the depreciation of the building and equipment, etc.) are considered period costs in Aunt Allee's Jams and Jellies accounting system.

Conrad explains that the indirect labor includes all the hours of the four-person production crew not charged as direct labor. They are all paid $12.50 per hour and work 2000 hours per year. Conrad and Edith agree that the indirect labor hours of the four-person crew is either setup labor or maintenance. The indirect labor is largely driven by the hours that the machines are in use, which is equivalent to the direct labor hours.

Table 52-3	Overhead

Indirect labor	$16,582
Fringe benefits	$50,000
Machine maintenance	$20,000
Cleaning & disposables	$20,000
Energy	$12,000
Indirect materials	$ 4,500
Scrap	$24,038
Office worker	$25,000
Total	$172,120

Fringe benefits are based on wages and amount to an additional 40% (insurance, FICA, benefits, etc). They can be allocated among the products based on direct labor hours. Machine maintenance costs are directly related to machine use, which can be tracked by the number of units produced. Cleaning supplies and disposable parts are used during the setup of the processing line. Usage is reasonably proportional to the number of setups (which is equal to the number of batches produced).

Energy costs are most directly related to the number of units produced. The scrap costs are directly related to setup, and usually amount to 5% of the value of materials.

Indirect materials are mainly shipping supplies. Their consumption is a based on orders shipped. As an order can be a mix of products and can be of varying quantities of product, Conrad does not see an easy cost driver. Since it is a relatively minor sum, he figures that sales volume is as good a method of allocation as any other.

Conrad estimates that the office workload is the same for cherry jelly, cherry jam, white grape jelly, and white grape jam. The workload (parts administration) caused by blueberry jam is 1.25 times as high as for any of the other products, due to supplier issues. The office worker makes the same pay rate as the production workers.

Based on the activity-based costs, what margin is each product achieving? Should Aunt Allee's Jams and Jellies expand their product line to include more low-volume specialty jams and jellies?

Problems in Pasta Land

by
Andrés Sousa-Poza
Old Dominion University

The Food Factory has been operating in an underdeveloped country for approximately 10 years.[1] Its parent corporation specializes in wheat milling, and it started the pasta factory as a "side-line" operation to process lower quality wheat flour, which is a by-product of the normal milling process. This low-gluten flour is generally not suitable for the production of bread or for direct sale to consumers.

In 2009, the pasta division is confronted with a major problem. It is too successful!

The factory was designed around the mill. Production capacities matched the amount of effluent from the mill rather than coming from a sound marketing strategy. As shown in Table 53-1, by 2006, the pasta plant was no longer able to effectively serve existing customers. The plant that was designed to produce 600 tons of pasta per month on two production lines is now facing average monthly orders of approximately 800 tons. Furthermore, the corporate director of marketing estimates that orders could easily be increased to 1400 to 1800 tons per month.

---

[1] All monies used in this case are in the local currency, which is one of the more than 40 countries in the world that use the $ symbol and most of which are called dollars.

Table 53-1   Average Monthly Orders/Production

Year	2000	2001	2002	2003	2004	2005	2006	2007	2008	2009
Orders	200	280	360	490	450	580	620	710	760	800
Production	200	270	365	500	440	575	590	610	580	570

Another challenge facing the factory is that the initial equipment was refurbished, not new, and it is now antiquated and seriously dilapidated. Unless the plant is shut down, equipment replacement is going to be required. The existing equipment was already a technological generation behind when it was bought. During the last 10 years a new generation of equipment has been developed based on high-temperature drying. The new technology is much more suited for use with low-quality (low-gluten) flour and semolina. New machinery is significantly more efficient. It requires fewer workers, has lower relative energy consumption, and produces less waste. The pasta plant still maintains a price lead through the low cost at which it is able to obtain raw materials from the corporate wheat mill, but this barely compensates for the plant's low efficiency.

The new technology is also enabling competitors to use low-quality, low-cost raw materials and still produce a reasonably high-quality end product. Ultimately, this means that the cost of higher quality pasta has dropped significantly in price, *and* the quality of the low-cost pasta is increasing significantly. The pasta factory's market is customers with a preference for low cost. Serious threats that the marketing director is possibly not including in the potential sales forecast include (1) the encroachment of traditionally high-quality producers into the low-cost markets, and (2) the increase in the quality expectations of customers that traditionally have been classified as "cost-conscious consumers." In general, the plant's cost advantage is no longer enough to secure a strong market niche. Only the most steadfast price-oriented consumers can be counted on as a stable market group.

Expanding the factory would mean that the mill would no longer be able to supply sufficient raw materials. Production above 600 tons per month would require the purchase of flour or semolina from the open market, at market prices. It is suggested that some of this increased cost could be dissipated through internal accounting practices, namely by charging the pasta division less for the use of the mill's low-quality flour. The mill's manager is strongly opposed to this, even though it has been done in the past. He argues that a further decrease in the costing of low-gluten semolina would seriously distort the mill's operating figures. He argues that the effluent flour should be sold to pig farmers, since they are willing to pay more than the pasta plant is now being charged.

While everyone (the board of directors and managers) concurs that something has to be done, little analysis has been done. The board of directors has traditionally made major investments based on a heuristic of keeping the initial capital costs low. The logic behind this is that it tends to reduce risk, since the time needed to achieve a net positive cash flow is shorter (short payback period). The general manager is not convinced that this is the best approach now. He feels that the threat introduced by new technologies, the deviation from the initial intent, the stance taken by the milling manager on costing, the cost and quality demands being placed by customers, as well as operational efficiency and effectiveness considerations, make it imperative that new technology and a new decision approach are used.

The general manager needs to develop a solid business case for each alternative. He needs to present these in a manner that would be accepted and understood by the board of directors for his recommendations to be considered. He is sure that this will require better financial metrics. The inclusion and consideration of so many "soft" and barely quantifiable variables is confounding him.

General Alternatives

At a very high level, there are not too many alternatives. Continuing the operation "as is" is not possible. While shutting down is possible, it is very undesirable. The use of the factory as a value-added function within the firm is still very positively viewed, and even though the mill manager is definitely not in favor of reducing the cost of the low-gluten semolina, at the organizational level there are considerable cost benefits. Furthermore, the factory provides jobs in a community where alternate employment is not possible. The corporation as a whole is the largest local employer, hiring approximately 40% of the eligible workers. This is the basis of the company's good reputation, and, possibly more importantly, the reason that no local taxes are being levied on the firm. The buildings to house the mill and the pasta factory are also being provided for free. The firm as a whole is better off given the benefits the pasta plant provides the mill—even if the pasta division runs at a mild loss.

Therefore, something must be done to replace the existing equipment. The main questions seem to lie in the technology and the organizational objectives. Associated with the technology alternatives are issues related to employee skills that would be required for the new production systems, customer expectations, and whether the expected performance promised by new technology could be achieved. Associated with the organizational goals are customer demands. The organizational intent for the pasta plant did not place the customer in a central position. Customers expect that, at the least, orders that are accepted will be adequately serviced. The factory's original intent to use the mill's effluent doesn't match customer demands that exceed the approximately 600 tons per month of low-grade semolina provided by the mill. Furthermore, customer expectations of pasta quality are changing drastically. Operating the pasta factory as a "side-line" operation is deterring the organization from making customer-centric decisions.

Technology and Equipment

Pasta is made with a process that has remained virtually unchanged for hundreds of years. What has changed over the centuries is the manufacturing technology. Pasta is made by mixing durum wheat semolina with water to achieve a moisture content of about 32%. Durum semolina is in essence coarse flour made from hard-grained wheat. It has a high gluten (protein) content, approximately 11% moisture, a distinctive "nutty taste," and an appealing yellow color. This mixture is kneaded and formed. Finally, the pasta is allowed to dry in drying rooms or in the outdoors.

In modern facilities, continuous feeding/dosing apparatus and paddle troughs accomplish mixing. Kneading and shaping is carried out using screw extruders and extrusion molds to give the pasta its characteristic shapes. Short pastas, such as elbow macaroni, penne, and fusilli, are cut short at the extruder. Long pastas, such as spaghetti and linguine, are draped over rods and cut to length once dry. The extruded pasta is dried in a series of dryers that bring the moisture content down to approximately 11.5% to 12%. The pasta is then cooled and stored before being packaged.

Figure 53-1   Short Pasta Production Line

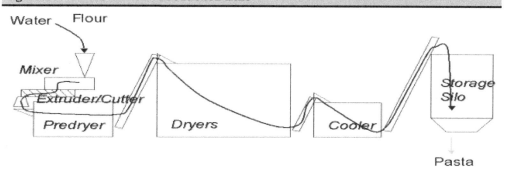

Possibly the greatest advances in the last decades have been made in the manner that pasta is dried. In older operations, the drying cycle can take 24 hours or more. This extended drying period is required to ensure that drying takes place evenly across the profile of the pasta. If drying takes place unevenly, stresses are caused in the pasta, causing it to crack and discolor. New technologies that rely on very well controlled drying environments use high temperatures that are moderated by adequate humidity levels. Drying time has been reduced to as little as 3 hours for specific pasta shapes. In addition, these high-temperature techniques can produce pasta with good coloration, texture, and cooking characteristics using flour with low gluten content (~10% or less), grayish color, and lacking the "nutty" taste of the durum wheat.

The use of flours with lower gluten content has a strong financial implication, since it is significantly less expensive than durum semolina. In most countries it is now common practice to include at least a portion of flour with the durum semolina. Many companies maintain at least a token amount of durum semolina content to be able to claim its use. In Italy this practice has been curtailed due to the "purity" laws that exist for pasta.

The general manager has found that there is a considerable difference in the prices of refurbished and new equipment. Refurbished equipment costs only 25% of the new machinery cost when transportation, installation, and start-up are included. As shown in Table 53-2, the price does not vary as much with the machinery's capacity.

Table 53-2	Equipment Costs		
		New	Refurbished
1000 tons/month		$3,560,000	$730,000
2000 tons/month		4,800,000	1,100,000

As shown in Table 53-3, new equipment, although it requires less frequent maintenance, uses parts and labor that are significantly more expensive. Refurbished equipment is more expensive to maintain as wear increases. However, power consumption is lower for the new equipment.

Table 53-3	Operating Costs		
Equipment	Maintenance costs		Power consumption/ton
New	7%	of equipment cost	$69
Refurbished	6%	of equipment cost	72
	& 1%	increase per year	

Labor and Skills

A factor of great concern to the general manager is the changes that new technology would bring to the organization. According to the equipment's manufacturers, the new technologies (which include advanced process control equipment, automated product testing, and integrated machinery testing) would require fewer employees than the refurbished equipment. The high-capacity equipment requires only a few more employees than the low-capacity machines, and the manufacturer argues that this is a good reason to implement a high-capacity facility.

As shown in Table 53-4, all options employ fewer than the 80 to 90 people that currently work in production. The existing equipment has very limited packaging machinery, and the

228

current process is very labor intensive, particularly for the first step where individual 500g boxes of pasta are put into case packs. Adjusting the labor force size to that required for the high-capacity refurbished technology could probably be accomplished by regular attrition. The other options would very likely require some layoffs. This concerns the general manager, who realizes that a strong component of the firm's good reputation is due to the number of local people that it employs.

Table 53-4   Labor Requirements by Equipment Type

| | | Capacity | |
Equipment	Skill Level	1000 tons/mo	2000 tons/mo
New	Supervisory	1	2
	Skilled	34	41
	Unskilled	2	8
Refurbished	Supervisory	1	2
	Skilled	2	5
	Unskilled	48	67

The new equipment also requires more *skilled* employees, who cost more (see Table 53-5). Even more important is that the local labor force does not have enough skilled employees. Hiring outsiders would exacerbate the loss of jobs for the local population. Extra training costs if locals are used are summarized in Table 53-6.

Table 53-5   Labor Costs per Month

Supervisory	$6700
Skilled	5300
Unskilled	3800
Note: Costs include benefits and training.	

Table 53-6	**Extra Training for New Technology**				
Year		1	2	3	4-10
Extra training as % of labor costs		22%	15%	8%	5%

The greatest problem in overhauling the factory, and in particular for the new equipment option, lies in the attitude and perceptions that exist within the factory—all the way from the employees to the board of directors. The following represent some of the basic attitudes that the general manager has encountered over time:

1. "If it isn't broke, why fix it?" Even initiatives aimed at improving conditions for employees have met with stiff resistance. For example, an attempt at starting a day-care facility failed. Management was not convinced that it would provide a tangible benefit. When several employees complained about the incursion on their privacy, the initiative was laid to rest.

2. "This is the way we have always done it." The installation of conveyor belts to move end product to the warehouse turned into an expensive fiasco. The employees refused to use it, and possibly even sabotaged equipment to support the claim that it was worthless. The employees seemed to fear that the workforce would be reduced.

3. "There is nothing that technology can do that a person cannot do better." When an attempt was made to introduce computers in the warehouse to log movement of incoming and outgoing materials, there was nearly an open revolt. Finally, the computers were removed from the warehouse, and the old paper system was reinstated. The loading dock supervisor would record everything on a clipboard and later enter it into the tracking spreadsheet.

4. "You cannot make cheap products on expensive equipment." It seemed that there was nearly a phobic fear of purchasing anything new. The usual excuse was that

it cost too much. In several cases where the general manager had been able to demonstrate (using sound financial and economic techniques) the benefits of adopting new equipment, other alternatives had been chosen.

5. "Training is for dummies." The Human Resource manager would complain that the only way he could get employees to attend even the most mandatory of training, such as training on hygiene techniques in the food industry, was through coercion. He would, in effect, threaten to have people fired.

6. "If the customer does not like the service, does not like what they get, does not like the quality, etc., let them find it cheaper somewhere else!"

Not surprisingly, concepts such as quality control, improvement, customer focus, hygiene, empowerment, participation, or, for that matter, customer and product are not really well understood.

Based on experience and research, the general manager also has found that introducing new equipment (particularly when new technologies are involved) requires a learning curve to realize the machinery's full capabilities (see Table 53-7).

Table 53-7	Learning Curves for Production (% of capacity)				
Year	1	2	3	4	5+
Refurbished	90%	95%	98%	100%	100%
New	45%	65%	90%	95%	100%

Customers

Existing orders (some unfilled) amount to approximately 800 tons per month. The director of marketing predicts the sales levels shown in Table 53-8 for the new equipment if not limited by the equipment's capacity or the learning curve.

Table 53-8	Forecast Sales for New Equipment			

		Year		
		1	2	3+
High sales target		1200	1600	1800
Low sales target		1200	1400	1400

The general manager feels that, under the present conditions, both the target range and market growth rates are feasible. He foresees some serious threats emanating from the new technologies being used. Based on past sales records, he finds that the company has six retail customers with which it has dealt on a regular basis. Of those six, one of the retailers has placed more than half of all orders (54%), with the second largest one placing another 30%, and the other four placing nearly equal portions of the remaining 16% of total orders. Independently, the general manager has talked to the customers and has found that both of the two most significant customers have plans for expanding sales, which bodes well for expanding the company's production.

These two customers are, however, quite specific in their expectations for service and indicate that they have been receiving offers from other suppliers who seemed to be in a better position to provide them with the required product. Furthermore, they indicate that consumers are beginning to show a preference for higher quality pasta. The largest customer in particular states that pasta quality would have to be increased; otherwise, they would have to seriously consider switching suppliers.

This makes it clear to the general manager that he has serious quality issues. Using the refurbished technology, he could only match competitors' quality if he uses a blend of high-quality semolina (which he would have to purchase from the open market) with the current low-quality flour. The blend would have to be approximately a 50:50 mix. This blend would obviously greatly dilute the beneficial effect of using very low-cost raw materials on the overall product costing. As shown in Table 53-9, semolina and flour are the greatest of the direct costs.

Table 53-9   Direct Costs per Ton of Product

Semolina from mill[a]	$770
Market low-grade semolina	940
Market high-grade semolina[b]	980
Other materials	128
Transportation	145

[a]A maximum of 600 tons is available.
[b]The refurbished technology requires a 50% inclusion of high-grade semolina.

The general manager is concerned with using a blend of flour and semolina since operators with the new technologies could do the same and have other advantages in market maneuvers. Since the retail customers were showing a strong preference for the product from the new technologies, he feels that he (and his competitors) could charge a small premium for the increase in product quality (see Table 53-10).

Table 53-10   Price of end product

Present price & price using refurbished technology	$1,450
Price of product using new technology	1,470

After discussing cost and quality issues in depth with the customers he feels that although the target values of 1400 to 1800 tons/month are reasonable, there is in effect a very high likelihood that he might lose at least one of the major customers if he uses the refurbished technology. If this occurs, the most likely scenario would be that he would have to revert to competing on a cost basis. The limitation on this is of course the amount of flour that he can obtain at a discounted price from the mill. Consequently, he feels that with the refurbished technology the likelihood of having a long-term sales outcome of approximately 600 tons/month is nearly as likely as the higher targets. Furthermore, there seems to be the danger of losing all the anticipated increase should the trend in pasta consumption shift further toward quality and away from price, although he feels that this is not quite as likely.

On the other hand, with the new technology it is much more likely that he can compete with other producers on a "level playing field" and achieve sales between 1200 and 1800 tons per month.

Financing

The finance director states that loans are available at 8% over 10 years to cover the full cost of equipment purchases. The director has used this leverage to its full extent. In general, projects hava been financed by paying 10% of the investment using retained earnings (although the director expresses reservations about exceeding $100,000 in this case).

The company pays taxes at approximately 35% of net earnings. Net losses for the project would reduce the firm's overall income and its tax burden. This type of equipment is depreciated to a salvage value of zero at 8 years with straight-line depreciation.

Finally, the director of finance informs the general manager that it is common for the firm to expect a minimum return of 20% on a project of this duration and scope, and that a 10-year planning horizon could be used. With this horizon it is reasonable to assume that the equipment's salvage value approximately equals the cost of removal.

The Problem

The general manager's overall impression is that the greater cost of new equipment will make the alternatives that included it more sensitive to fluctuations in the eventual sales. On the other hand, the refurbished technologies risk failing to meet rising quality expectations. This could easily curtail any sales growth and even put this plant out of business altogether.

He also sees that no matter what the outcome, he is going to have serious labor issues to deal with, and a lot of discussion with the board of directors to determine the future focus of the pasta plant. What are your recommendations?

Whirlwind Exploration Company

by
Michael Dunn
Petrotechnical Resources of Alaska

Whirlwind Exploration Company has identified northern Alaska as a prospect for oil and gas exploration. The region is considered a world class petroleum system, and it is famous for its huge discovered oilfields and vast expanses of under-explored land. The region is also known for its high cost structure and relatively low wellhead oil price due to distance to market.

Oil and gas exploration is inherently risky. In exploration areas like northern Alaska, a geologist is lucky if he finds producible oil in one out of four wildcat wells. To make matters worse, only one in three geologic successes may be large enough to be a commercial development. Given these odds, before committing to an exploration program Whirlwind must carefully compare the area with other investment opportunities.

Whirlwind has studied the geology, noted the field sizes of existing fields, and gathered data about wellhead price and development costs. As a first step to determine whether Alaska is worthy of investment, Whirlwind must calculate how large a field must be in order to break even. This breakeven threshold size is often referred to as the "minimum economic size."

The geoscientists have mapped a prospect that is 100 miles from a producing oil field. The seismic data and well information of nearby wells indicates that the range of reserves can be described with a log-normal distribution. Statistical data from fields all over the world have determined that the three components that define reserve size—area, net pay ($\equiv$ average formation thickness), and recovery factor—can also be described with log-normal distributions. Using the available seismic data, net pay of nearby wells, and recovery factors of analogous producing fields, a Monte-Carlo simulation has been completed. This has resulted Table 54-1, which defines a "reasonable" combination of area, net pay, and recovery factor for a range of reserve sizes.

Table 54-1	Reserve Size, Area, Net Pay, and Recovery Factors		
Reserves (bbl)	Area (acres)	Net Pay (feet)	Recovery (bbl/acre-ft)
100,000,000	6,050	65	256
200,000,000	8,800	77	295
300,000,000	11,000	85	320
400,000,000	12,600	91	350
500,000,000	14,100	96	370
600,000,000	15,300	102	385

The reservoir engineers have determined that producing wells for the field should be spaced at one well per 160 acres to recover the reserves in the optimal time frame. They have also provided Table 54-2, which describes the field's production profile. The drilling engineer has determined that the wells will cost $8 million per well to drill and complete. He has assumed that one-third of the wells will be drilled before production start-up and two-thirds of the wells will be drilled following start-up.

Table 54-2	Production Profile				
Year	Rate (% of Reserves)	Year	Rate (% of Reserves)	Year	Rate (% of Reserves)
1	6%	6	10%	11	5%
2	9%	7	9%	12	4%
3	10%	8	8%	13	3%
4	10%	9	7%	14	2%
5	10%	10	6%	15	1%

The facility engineer has determined that the cost of the process plant and infield pipelines is a linear function of peak oil rate with a substantial fixed cost for gravel and logistics. The equation for facilities costs is:

$$\text{Facility Capital Costs} = 100,000,000 + 8,000 * POR$$
$$\text{where } POR = \text{Peak oil rate (barrels per day)}$$

The cost of a sales line to transport sales quality crude to the metering station at pump station 1 of the Great Northern Pipeline has been estimated to be $4 million per mile. The operations department has determined that the operating costs are $12 per barrel. The oil marketing specialist has estimated the sales price of crude at pump station 1 at $45 per barrel, flat real, for the field's life. The finance department has determined that a 25% interest rate should be used to evaluate this prospect.

As summarized in Table 54-3, the engineers and permitting specialists have determined that the facilities and pipeline capital will be spent over four years. The operations department has estimated that abandonment costs will be equal to 20% of facility and pipeline capital costs, in the year after the field is shut-in.

Table 54-3   Start-up and Shut-down Summary

Year	Facility & Pipeline Capital Spending	Rate Profile (% of reserves)	Drilling Capital Spending
1	15%		
2	20%		
3	35%		
4	30%		1/3 of wells
5	Production startup	6%	1/3 of wells
6		9%	1/3 of wells
7		10%	
19		1%	
20	Abandonment costs	0%	

Given this information about the potential size of the field, its relationship to size and area, the costs to develop and operate the field, and the sales price of crude oil, what is the minimum economic size of a field at his location?

The simulation of the range of reserves produced a low-side reserve size of 133 million barrels (estimated probability of being exceeded of 90% or P90%), a median estimate of 266 million barrels, and a high-side reserve size (P10%) of 500 million barrels of recoverable oil. What is the approximate probability of achieving the minimum economic size?

Suggestions to the Student

1. Pick a reserve size (say 300 million barrels of oil).
2. From this determine the peak *daily* oil rate and the facility and pipeline costs.
3. Determine the number of wells for the area associated with the reserve size, and then calculate the drilling costs.
4. Build a cash flow table and calculate the PW and IRR.
5. Pick a second reserve size on the opposite side of PW = 0. Interpolate to find the minimum economic size.

Printed in the United States
By Bookmasters